# PALMS & CYCADS

## AROUND THE WORLD

# PALMS & CYCADS AROUND THE WORLD

*Jack Krempin*

KREMPIN
BOOKS

Distributed by
Krempin Books
25 Beverley Crescent
Broadbeach Waters
Queensland 4218, Australia.


National Library of Australia Card No. and
ISBN 0-646-14252-6



Printed in Hong Kong by
Excel Graphics Art Company.

# FOREWORD

*Palms & Cycads Around the World* takes a comprehensive look at the world's palms and cycads, with some look-alikes often confused with both species. It is the first book of its kind to comprehensively feature a large range of the world's cycads in this way. At last, gardeners are able to find all these plants of common interest within one cover where 150 varieties of the world's 200 plus cycad species and 400 of the world's most popularly grown palms are described and pictured, many having been photographed in the wild. It also features the latest botanical names from our botanists, along with previously known names and many of the common used. This should dispense with much of the confusion existing with naming and identification. With this most comprehensive book of palms and cycads from around the world accompanied by the number of species of each, readers will find a treasury of up-to-date information, couched in simple terms, on all aspects of growing in the garden or in pots both outdoors and indoors — illuminated by a host of illustrations. The result is a reference and guide relating to matters like origin, world listing of species, sex for seed production, propagation, culture and natural habitat. Growing by seeds or division, compost mixes, potting techniques, feeding, watering, pest control and trouble-shooting are among other areas covered.

Botanists, who always seem to rely on the Latin language when writing, might frown at the simplified descriptions, but botanical expressions have been deliberately put into plain language as the Latin terms only serve to confuse 99.9 per cent of most gardeners. The addition of photographs provides a further recognition to these written descriptions.

*Palms & Cycads Around the World* contains details and pictures of more than 400 of the world's most popular and sought after palms as well as information about all the known species including the number of varieties of each. It includes a quick reference listing, covering 23 specific purposes such as climatic zones, size or type, single or multiple trunks, feather or fan leaves, wet or dry areas, sun or shade, growing tolerances, and so on, to cover most aspects of use and need.

The information on varieties is so extensive, the book should still be the most up-to-date market reference well into the next century.

*Palms & Cycads Around the World* devotes a small section to a score of palm and cycad look-alikes, but does not examine all species, discussing only the most common because they are not essential to the book's subject matter.

*Carpentaria acuminata*
A handsome row-planting for the tropics.

*Phoenix canariensis*
Hardy palm for all but icy regions.

*Cycas media*
About 200 years old.

*Beaucarnea recurvata*
Nolina Palm or Pony Tail. A palm look-alike.

The interest in cycads has increased dramatically in the last few years along with that in palms. Both have varieties in danger of extinction. With palms this could be occurring by human expansion into their natural habitats or preferred variety selection and neglect of the remainder. With cycads the same may apply also through rural, urban and industrial developments. It is a tragedy to now endanger many of what are claimed to be one of the world's earliest forms of plant life. Fossil finds have dated cycads back 180 million years. In rural areas, cycads are under threat from graziers whose hungry stock may be tempted to eat their poisonous foliage. This endangered status has created a demand by collectors and cycad enthusiasts to extend their collections before it is too late.

Past publications on cycads have been rare and brief. *Palms & Cycads Around the World* covers all known species from collectors' lists and incorporates most of those introduced into horticultural collections, botanical gardens and nurseries.

Cycads not pictured were beyond reach of our camera, in inaccessible wilderness or unfriendly countries. But there are still over 200 cycad pictures — many of them aged giants. To achieve this the author travelled the world, capturing on film, specimens from 20 to 200 years old.

Global interest in these ancient plants is intensifying and Australia is richer in cycad varieties than most other countries because of its vast wilderness areas. No one knows the Australian cycad better than Len Butt, who has added his expertise on the subject to this book.

A bonus to the listings of all known species of cycads is Len Butt's article, 'The Cycads of Australia', which has significance to the growing of cycads in other parts of the world. Don't miss reading it.

Jack Krempin

# CONTENTS

# REFERENCES


Blombery A. and Rodd T. 1982. *Palms.* Angus & Robertson. Australia.

Giddy C. 1974. *Cycads of South Africa.* Purnell, Capetown, South Africa.

Graf A.B. 1978. *Tropica.* Roehrs, East Rutherford, N.J., U.S.A.

Graf A.B. 1982. Series 4 *Tropica.* Roehrs, East Rutherford, N.J., U.S.A.

Jones D. 1984. *Palms in Australia.* Reed, Sydney, Australia.

Langlois A.C. 1976. *Supplement to Palms of the World.* The University Presses of Florida, Gainesville, U.S.A.

McCurrach J.C. 1960. *Palms of the World.* Harper and Brothers, New York, U.S.A.

Money D.C., 1970. *Basic Geography. University Tutorial Press, London.*

Stevenson G.B. 1974. *Palms of South Florida.* Self published.

Uhl N.W. and Dransfield J. 1987. *Genera Palmarum.* Allen Press, Lawrence, Kansas, USA.

# ACKNOWLEDGEMENTS

Firstly I lovingly thank my wife Eileen, who has once again helped me with the typing and had to patiently forgo outings and company while this was completed.

Then I offer my gratitude to all those who have courteously assisted me with their time, with information, with assistance in picture taking, or in just allowing me to use their facilities or plants. My special thanks go to the following:

- John and Phyllis Lee, for this assistance and patience;
- Ron McCahon, my bastion of strength;
- Dusan Balint, for editing my Cairns photographs;
- John Covacovich and Ray Groch, for their assistance;
- Len Dellow, for giving me the opportunity to work amongst palms, and sharing his long-term knowledge;
- Steve Sharkey and Harold Caulfield, for assistance with my Brisbane slides;
- John Forlonge, for assistance with my Sydney slides;
- Stan Walkley, whose wonderful nursery afforded many cycad pictures;
- Len Butt, for willingly sharing his cycad knowledge, which is unmatched; also for his article and the many pictures by him and his friends used in this book;
- Don Stallard, the Northern Territory wanderer, for his wonderful habitat pictures;
- Chris Migliaccio, my US acquaintance, for his many pictures;
- The staffs of the botanic gardens in Cairns, Brisbane, Sydney, Singapore, and Honolulu, and of Huntington Gardens California and Fairchild Gardens in Miami. My thanks to you all, for your care and effort in making and keeping those places the wonderful achievements they are, and in particular to Chuck Hubbuch for all his help, and finally to
- Vincent Hampson, who edited my manuscript.

# INTRODUCING PALMS & CYCADS

To successfully grow palms and cycads in your own garden it would be wise to consider the origin of each before purchasing them: Losses could occur if the climate differs too much from that of the plants origins, minor tolerances may only need minor adjustment in the garden situation.

The greatest majority of the palms and cycads originate from equatorial, tropical, subtropical and temperate areas between latitudes 40° north and south of the equator. There are a few exceptions either side of these latitudes dictated climatically, as well as a few varieties of palms and cycads that can even survive in snowy regions. But they are in the minority with the majority of either species unable to tolerate heavy frosts or snow in the open garden or natural habitat. A certain number originate from high tropical altitudes, giving them a degree of tolerance to some cold, but many come from equatorial rainforests and lack such endurance.

Others have their roots in places as varied as desert oases, tropical or warm temperate rainforests, open areas, along the banks of streams, in swamps, even soak areas in arid regions. These should have the capacities to bear hot or cooler dry-air climates and all have their place in a particular horticultural use. That is why it is important to understand their sources for success.

The book supplies the origin of each palm or cycad along with the natural growing habitats. It is worth considering these and trying to marry such knowledge into the after-care of that palm or cycad.

The following world maps will indicate the distribution of both palms and cycads in their natural habitats. It is evident by looking at the maps, that both species in nature are located within the latitudes between 40° north and south of the equator.

*Archontophoenix cunninghamiana*
Piccabeen or Bangalow Palm. With vegetation in natural habitat.

## THE ORIGIN OF PALMS

Palms are relative newcomers on Earth, according to fossil remains. These date Sabal palms as the earliest known species, at about 85 million years. Even today Sabals are still found growing under the toughest of conditions, in hot barren areas.

*Sabal palmetto*
Sabals are one of the oldest palm species, from fossil records.

## PALMS IN NATURAL HABITAT

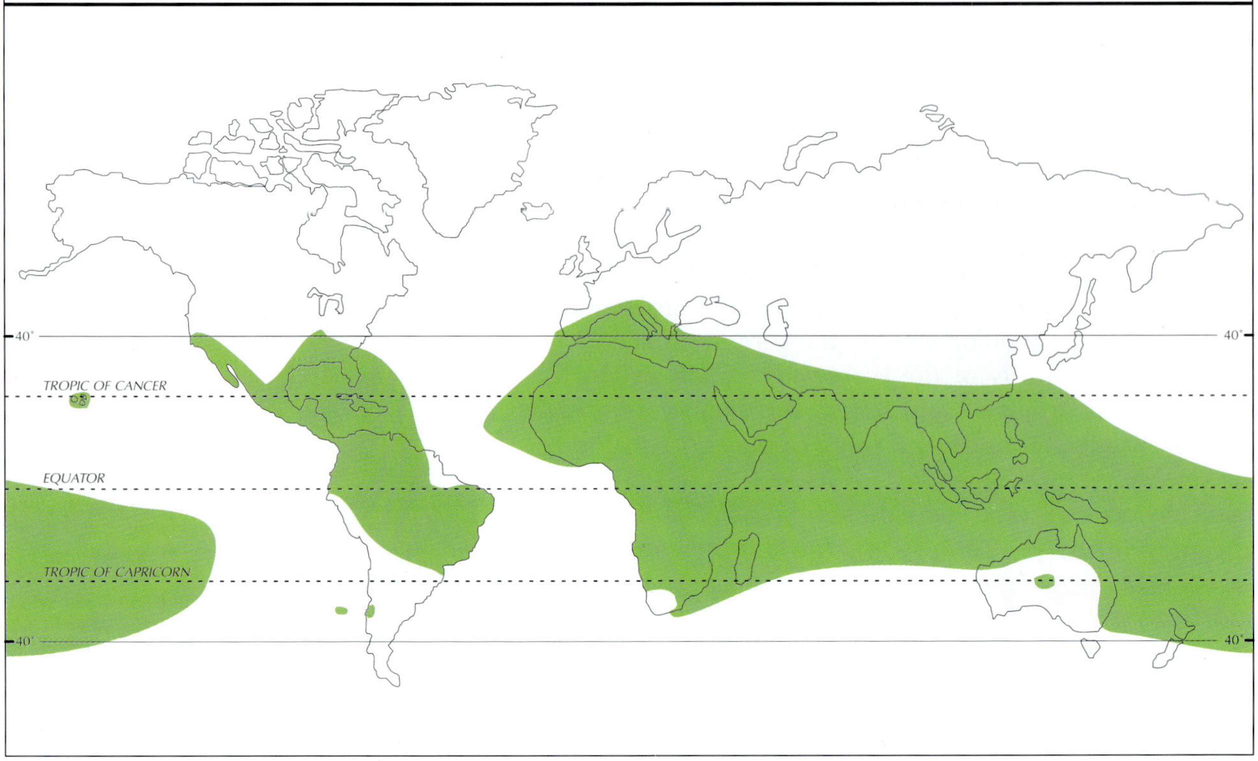


## CYCADS IN NATURAL HABITAT

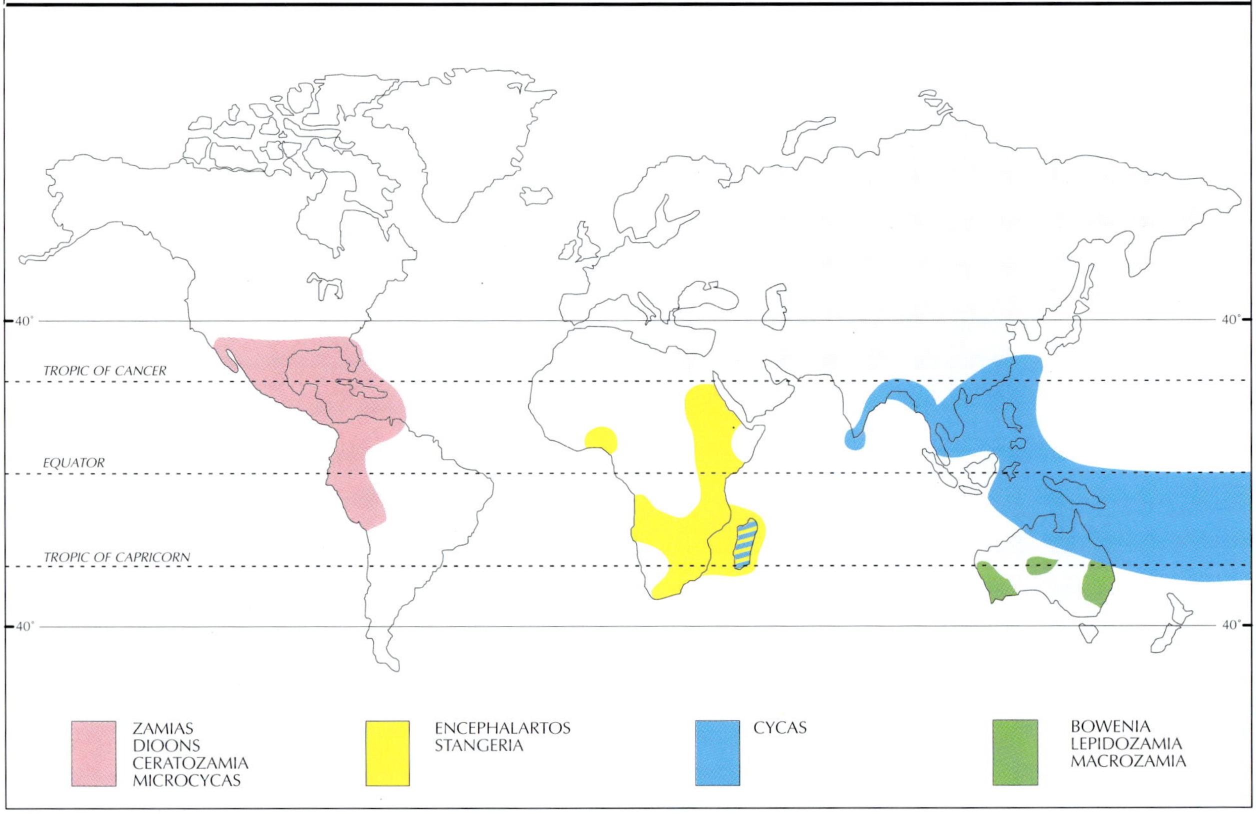

Many species have evolved in those 85 million years. With the environment changes from rugged, turbulent, volcanic landscapes to humid, tropical jungles. Most palms however, are located in lusher areas of high rainfall, though many are still around rocky sea cliffs, on sandy seashores, sharp ridges at high altitude, or even in deserts near water.

Palms have adapted well on this earth, with about 2800 known species in the 200 genera. These figures change from time to time as our botanists research the species and reshuffle the varieties to where they now rightfully belong.

Palms have travelled the earth in many ways. Some, like the coconut, float over the oceans to distant shores moving further afield with each generation, until eventually established on every tropical coastline. Other palm seeds have found accommodation in the stomachs of birds and animals in order to begin life further afield from their parent. Exported as fruit (dates) they end up on foreign shores and establish there, when the seed is discarded. And as highly prized garden and indoor plants, the seeds travel from country to country in the hands of collectors and commercial growers.

Palm species are spreading quicker than ever before in our modern fast-moving world. But as travel shrinks the borders, there is the threat that expansion may engulf and swallow up our nature areas.

It is to be hoped that all species survive and native varieties are not destroyed or replaced in preference by newcomers from overseas. Our conservationists will have to be alert to ensure this does not happen and prevent native varieties being lost forever. Vegetation's worst enemy is humankind, as we spread our way across the earth, destroying natural growth to farm the lands, establish grazing, build homes and cement cities. It is to be hoped that we can at least retain intact some areas of nature. It is a shame that survival cannot be planned ahead of expediency and greed, thus eliminating total destruction of what nature has provided. But humans threatens the environment, perhaps even life itself, with our evil inventions and weapons of destruction.

Already, a number of the world's loveliest palms have been lost to the wilderness and now only exist as horticulturally grown species. As more and more enter onto the trendy marketplace, many endangered species could become extinct. The decline of stocks in native habitats shows that our popular botanical gardens and reserves should be further extended to all areas of the globe, to fill these needs, before it is too late.

The origins have been recorded with the text of each species. The zone origins in the quick reference chart are only general for that palm, not the individual varieties.

*Butia capitata*
Wine Palm or Jelly Palm of South America.

*Veitchia winin*
A tall and stately palm from the New Hebrides.

*Cycas angulata*
Very ancient plants from Australia's Northern Territory.
(Photo D. Stallard)

*Macrozamia macdonnellii*
A cycad from Australia's hot inland ranges.
(Photo D. Stallard)

## THE ORIGIN OF CYCADS

Cycads are not related to palms, and the species are at least twice as old according to fossil datings. Their age has been established as about 170–180 million years, making them an ancient, enduring form of land vegetation, but there was life in the sea much earlier than this. Millions of years later, cycads started forming flower cones, beginning their evolution of the species through seeds. However, despite the antiquity of the species, collector-interest in them is only now gaining momentum, with gardeners beginning to collect the endemic plants, as well as the exotic ones from all around the world.

It will be noted from the accompanying map, that most cycads now inhabit the less-productive land areas. In this they differ from many of the palms, which occupy richer soils needed for crops, or grazing.

Cycads have adapted to a wide variety of environments. They can be found in dry, arid areas, which are open or have light tree cover, but they are equally happy near sea coasts or inland, in woodlands, rainforests, lowlands, highlands, among sandy scrub or even ancient volcanic gorges.

Some of this country could be required for grazing purposes. This is evident in both Australia and to a lesser degree in Africa, where the country is being grazed by cattle and sheep.

The magnificent Australian *Cycas* plants, many hundreds of years old, are endangered because of their poison threat to cattle when they are in juvenile new foliage. Stock tend to eat them in times of drought or after bushfires when there is a subsequent fodder shortage. As the foliage stiffens with maturity, the danger minimises because the plants no longer appeal to hungry animals. Native animals have learned to leave cycads alone, which is perhaps one reason why cycads have survived for 170–180 million years. Some animals do eat the outer fruit, carrying the seed to further destinations.

Indigenous people who have lived with cycads during their long existence on the planet, have found ways of overcoming the plants' poisonous properties to use them as food. However, European settlers are not so tolerant, they only finds ways to destroy these beautiful old specimens and their progeny, from what is now *their* grazing lands. I wonder if this is the way God intended it.

The map indicates the distribution of cycad species, by showing them in different colours. It is interesting to observe that each continent has its own range of varieties and these are rarely repeated on the next continent. The one exception is *Cycas*, which is present in the Asian, South-East Asian, North Australian and Pacific areas.

As with palms, it will be in the interest of the grower of cycads, to consult the origins before selection of varieties. Avoid unnatural extremes.

## PALMS ARE IMPORTANT

Palms have been of importance to humans throughout the ages. They have supplied many primitive needs, with trunks a source of timber for buildings and canoes, or for food (sago) and the slender ones used for spears. The leaves have been thatched and woven for a multitude of purposes, while the flower spikes have produced sugar for sweets and alcoholic drinks, and the fruits were either eaten fresh, preserved (dates), or used as a poison (*Orania*). We all know the uses for coconuts and more recently, the uses for palm oil in food products and oils for industry. There are also palms which supply wax, palms with narcotic properties (Betel nut) and palms with medicinal uses. In all, the palm has played, and continues to play, a very important role in people's lives — primitive and modern.

In many countries, palms are only commercially important to nurseryworkers in outdoor landscapes and indoor decorations for homes and public buildings — a growing industry in today's world. Up to about 20 years ago the nurseryworkers only grew a very limited range of palm varieties. However, enlivened interest brought about by the demand for different and better indoor plants, both for commercial hire and home decoration has seen the growing of palms skyrocket. This in turn has led to experiments with newer varieties, most of which come from undercover positions in tropical jungles, or out of warm temperate rainforests to satisfy this indoor need. New varieties find their way onto the market every year, as supplies of seeds trickle through from distant areas.

Another major reason for the popularity of palms is the home swimming pool. Nothing supplies that cool, relaxed appearance better than a palm. They are clean growers and do not continually shed small leaves as most trees do, and the leaves which are shed are large and can be carried away and disposed of easily. Some palm varieties produce an abundance of fruits (*Archontophoenix, Phoenix* and *Syagrus*), which are an attractive feature of the tree. It is probably better to leave these in place as long as possible. If they are not in the way there is no problem, but if they are the bunches of fruit may be removed at any stage before ripening, simply by cutting the whole bunch off. This avoids the untidy mess which arises when seeds begin dropping off around or in the pool. If palms overhang the pool it would be advisable to remove the inflorescence, the flower spike, before it opens.

*Cocos nucifera*
Coconuts have floated on the oceans to grow on distant tropical shores.

*Caryota urens*
Fishtail palms supply sago and sugar products.

*Cycas revoluta*
Japanese Sago Palm is the most popular cycad in cultivation. Male plant at top, female below.

*Encephalartos altensteini* (female)
Cycads have male or female cones in place of flowers.

## CYCADS ARE GAINING POPULARITY

For 170–180 million years cycads have graced both the more barren, less-lush areas of the world through to tropical rainforests. Perhaps, it may be said, these were up until now less populated areas of the world. Probably because of this and the ready availability of alternative foods to the indigenous peoples, cycads do not appear to have been valued as highly as palms.

There is only sparse recorded information available on their reliance as a source of food, but I daresay there would be far more in early folklore, if this were recorded. It is known that cycads have been used as a source of food. There is data on the crushing of the seed kernels into a paste, followed by lengthy washing of the pulp to remove the poison before cooking. The pulp was then spread on firestones to bake into a bread. There is no information available about whether this was a regular food or only a bushwalk snack, for Australian Aborigines. It is known that the African peoples also used cycads as food. *Encephalartos* have many common names linking them with sago and "bread palm". The Chinese and Japanese have also been associated by the name of *Cycas*, one being known as "Japanese sago palm". I daresay the early Incas and Indians of the Americas also had their use of this plant. There must be many interesting early uses for cycads if someone were to research the subject, or publish the already known data from various countries.

The plants are now proving increasingly popular with collectors and home gardeners for display purposes. It is not suggested that these people experiment with them as food, if they wish to avoid a tummy ache, through improper removing of the poison. Early Australian history records this misadventure. Cycads are singly sexed, so to produce fertile seeds, two or more of a species, one of either sex are required for fertile seeds. A married pair have more interest and can be rewarding. Cycads are slow-growing, live a long time, are manageable by not outgrowing, are hardy and mostly trouble-free. They are well-worth a place in a permanent garden, or a trial as an indoor specimen.

# PALMS

## PALMS AS GARDEN SPECIMENS

The planting of a palm, in preference to a tree, alongside a home, could also ease the leaves-in-the-gutter problem, which continually arises. Any palm fruits overhanging the guttering should of course be removed before they start dropping. In selecting palms for this purpose it may be wise to choose from the range of clustering palms, as well as solitary ones. Palms are wonderful cohabiters and can be mixed with their own varieties, as well as with other shrubs and trees; including ferns and some of the hardier plants from the indoor range. A further advantage which palms have over other trees, is that quite large specimens may be safely transferred for landscaping projects. It's a practice which has been used often in holiday areas and has transformed building sites into attractive holiday units giving them the appearance of having been in operation a long time.

Palms have a relatively small root growth for their size, but retain a sturdiness against quite strong winds. This allows them to be planted in rather confined spaces, that would be too cramping for trees of similar heights. Their roots do not spread like trees, but tend to mass very compactly, which is the reason they can be lifted and moved so successfully. Their growth is upright, particularly in the single-trunked species, giving an umbrella-like canopy for the planting of showy base plants. A filling effect may be achieved by using varieties which have multiple or clumping trunks. These also grow upwards, giving height, but continue to clump with new trunks thus keeping the base bushy.

Through the text on palm varieties, I will be constantly referring to suntrap and rainforest-type plantings. The previous chapter opened the door to this subject, so I will give you some ideas on setting these up further on.

Suntraps hardly need explanation, but in referring to them as a protection for palms, my meaning may be a little different to yours. Most people would probably think of an enclosed area, fully in the sun — perhaps where they could sunbake. In some cases, that would apply, but not all because varieties have however, differing needs. Some palms won't tolerate full sun, so in such instances a more correct definition would be a shaded suntrap. One which still had all the warmth, but was shaded from the intense rays of the sun, at its hottest, perhaps by a tree. The environment for tropical no-sun palms, would have to be warm and protected from cold winds, whilst out of direct sunlight.

*Roystonea regia*
Cuban Royal Palm enhances buildings.

*Chrysalidocarpus lutescens*
Golden Cane Palm has multiple trunks.

*Hyophorbe verschaffeltii*
Spindle Palm has a lovely crownshaft.

*Livistona decipiens*
A weeping fan palm, lacks a crownshaft.

## PALM FEATURES

With the different species of palms, there are a number of different leaf shapes. For simplicity I will be referring to these as either *feather palms* (botanically pinnate) which are displayed in a feather shape. *Fan palms* (palmate) are displayed like the palm of a hand or circular but divided by segments. *Fishtail* or caryotoid (bipinnate) are usually displayed feather-like with unusual shaped leaflets on branched stems. *Simple* or undivided leaves may remain whole unless divided feather-like by the wind. This is perhaps over-simplifying the various leaf shapes, as there are still more less-common variations.

I will constantly refer to certain palms having crownshafts. This is the section above the trunk and before the leaves, but is not present on all palms. Some crownshafts can vary in size, shape and colour, and add a clean look to the appearance of the palm. Also, all palms with crownshafts are self-cleaning — that is their old spent leaves drop off the palm trunks by themselves, always leaving the trunk looking clean and tidy. Usually, the inflorescence, or flower and seed spikes, are borne at the top of the trunk, just below the crownshaft adding a lovely "skirt" effect to some, as in Alexandra Palms. This does not mean palms which have crownshafts are the only ones worth growing. That would be far from the truth. Think of the beautiful *Syagrus, Livistonas, Caryotas* and *Phoenix,* to name but a few.

Instead of crownshafts many varieties of palms have their leaf stems coming directly away from the trunk. Some leave a ringed appearance on the trunk, once the spent leaves have fallen, as with crownshafts while others are shrouded in fibres and husks, and old leaf bases. Some do not shed these at all. Others shed them at various stages of trunk growth and height. Some palms have clumping or branching trunks. Some don't have any trunks at all, not visibly anyway. These trunks may be subterranean, either single or branching. Then there are trunks armed with spikes, which can be short thick and vicious or resemble large darning needles and perhaps could have been used for that purpose in primitive times. A number of interesting palms support their trunks out of the soil on stilt-like roots. In some cases these could be up to a couple of metres high. The differences in the varieties can be quite fascinating.

Not quite as fascinating though as the sex life of palms. With some, pollination is a very easy and efficient operation for the bees or other insects or animals. I have often gazed upwards as a beautiful spray of either *Archontophoenix* or *Syagrus* flowers unfold and hang in their glory. This is followed by the frenzied working of the bees and other insects, with a continual shedding of the spent sections, until the whole operation is ably completed. The next stage is very heavy bunches of fruits, which on the *Syagrus,* much resembles an enormous bunch of grapes, eventually dropping to resemble ripe loquats lying on the ground. With the *Archontophoenix* the bunch is broader, more open and spreading, and covered in green pea-like fruits, which mature to a shiny red. These, and numerous other varieties of palms, have both male and female sprays of flowers in the same inflorescence, making the task of impregnation for the insects a simple one. This is known as *monoecious* and is self pollinating. Other varieties carry bisexual flowers and these are known as being *hermaphrodite.* In further cases some palms are both monoecious and hermaphrodite and also self pollinating.

For the third classification nature has made the impregnation process much more difficult. These are *dioecious,* and as with people, the palms are male or female sexed, with the male flowers on one plant and the female flowers on another plant. As plants are stationary, they rely on insects, breezes or animals to carry the pollen from plant to plant. Palms like the *Phoenix* etc., produce an abundance of pollen which on still mornings, can be seen like a cloud or mist around the plants. This process must also be highly efficient because the seed bunches on *Phoenix* are always very full.

With some palms the reproductive process can be terminal. The *Corypha* a solitary trunk palm, grows its whole life, only to die after producing one outstanding head of flowers and seeds. Others will continue producing fruits on the trunk of the palm, or if multiple-trunked, on one trunk, until the point is reached, when that trunk, or that solitary palm, dies out altogether. This is evident with *Caryotas* and *Arengas,* but this point may not be reached for at least 15 years or much longer, so don't be put off by it.

The fruit sizes, shapes and coverings vary considerably. A good comparison is between *Archontophoenix* and *Cocos* (coconuts). The fruit on the former is about 1cm (3/8") in size and covered in a red flesh. This contrasts markedly with the coconut, about 15cm (6") in size, having a hollow centre full of rich liquid and covered with a fibre coating, that extends its length to at least 30cm (12"). Further interesting and unusual fruits are those of the *Astrocaryum,* and in particular *A. mexicanum.* The fruits are about 3cm (1"–1¼") completely covered in spines and too thorny to handle. More thorns can be found on the pouch-like coverings enclosing the fruits which are edible, and have a flavour resembling coconuts.

The purpose of each of these fruits, or other coverings,

Flowers of (A) *Syagrus,* (B) *Phoenix* (C) *Rhopalostylis* (D) *Archontophoenix*

Fruits of (E) *Syagrus* (F) *Cocos* (G) *Sabal* (H) *Archontophoenix.*

*Dioon edule*
Virgin's Palm from Mexico.

*Encephalartos lehmannii*
Blue-leaved cycad with a female cone and seeds.

is for the distribution of the seeds. Those with fruit are eaten by birds and animals, so that the seeds may be carried and distributed to another area, either inside or outside of that animal. Coconuts have a wonderful floating means of transportation, with a buoyant husk around the seed. They have travelled the oceans of the world, island to island, continent to continent, for thousands of kilometres until they are present on every tropical coastline. Coconuts have created export economies, supplied building timber and thatching, and stabilised food supplies for many primitive people.

# CYCADS

## WHY CYCADS?

These palm-like plants are collectively known as cycads, being a shortening of the collective name Cycadales and they include the botanical species of *Cycadaceae (Cycas), Stangeriaceae (Stangeria)* and *Zamiaceae (Bowenia, Ceratozamia, Dioon, Encephalartos, Lepidozamia, Macrozamia, Microcycas* and *Zamia)*. They are also commonly known as "*Cycas* Palms", "*Sago* Palms", "*Zamia* Palms", "Bread Palm" and numerous other regional common names.

My original aim when writing this book was to concentrate mainly on popular palms and the latest information available about them. In deciding to add "palm look-alikes" to my palm book, Cycads were to be included, but only a few species were to appear in that section. However, my contact with Palm and Cycad Society members, Botanical Gardens and progressive nurseries, showed me that this "Rip van Winkle" plant was really waking up with a vengeance. I found that most palm enthusiasts were also cycad enthusiasts, who wished to increase their collection of both native and exotic varieties. These are still not plentiful anywhere, as growth is very slow and the available varieties limited, but very much on the increase and limited only by availability.

Now, with the world interchange of species gaining momentum through these societies and nurseries, I would suggest cycads as the "Plant of the 90s" — even the next "craze" in plant popularity.

I decided to make cycads a major feature of this book, with palms, the other associated interest. In my effort to research cycads for added details of exotic varieties, I discovered there was little to no printed material available on the subject and nothing at all on any species other than *Encephalartos*, (a South African species). In the brief time available to me, before going to press, I have included as much information on each, as is accurate, leaving out any speculation. Many of the species are located in politically dangerous countries, making their inclusion and more comprehensive information impossible at present. Despite these setbacks, there are 150 of the world's 220 plus species recorded in cycad society files, nurseries and botanical gardens and with all the cycad families represented. The names of the remainder not pictured are also given, making a known world listing.

At present the cycads of the world are naturally collected species of those countries, about 220 in all. This may change in years to come, as man starts interfering, by cross-breeding among the species and varieties. There are signs of this already, in the assembling of collections, as only a single specimen plant, either male or female, represents a variety. Being singly sexed, either male or female, crossing will most likely take place, where a male of one and a female of another produce their cones at the same time. Deliberate crosses, for colour and form, will also probably be explored for profit and prestige.

*Stangeria eriopus*
Hottentot's Head, a fern-like cycad from South Africa.

## CYCAD FEATURES

In appearance many of the cycad species are similar to many of the palms, but they are unrelated. While they may have a trunk or trunks and be topped with an array of feather leaves, that is where the similarity ends.

Most do have an array of feather-shaped leaves, but instead of being soft, these are usually stiff, with sharply pointed tip spikes. A number of cycads have varying amounts of thorns along the edges of the leaflets, making them vicious plants to handle; a few species have firm leaflets, which are not spined at all, while other varieties sport a soft foliage. There are cycads with narrow leaflets and cycads with leaflets that are broader, and though leaf fronds do follow the pinnate of feathered shape, with some, this may only consist of a few leaflets.

Trunk features among the species are very diverse, with some (*Lepidozamia*) recorded at heights of 20m (65′). But trunk sizes vary according to the variety, with certain types being hidden underground (subterranean) or only displaying tuberous growths. Other trunks may be branched or clustered and can grow as thick as 1m (3′) or as thin as just a few cms.

*Zamia furfuracea*
Jamaican Sago Tree is a popular cycad.

On some cycads the leafstems are very stiff, while others are softer, even floppy. They can be straight as arrows or

*Stangeria eriopus*
showing the F cone

*Encephalartos ferox*
Bears a large attractive F cone

recurved with others, and have textures ranging from satin smooth to needle-sharp as they bristle with thorns of varying sizes.

Cycads differ from other plants in that they don't bear flowers to produce their seeds. Instead they bear cones, the M (male) cone/s on one plant and the F (female) or seed cone/s on another plant. Once pollen has fallen from the male cone this dies and drops off. With the seed cone this isn't so, the seed nuts continue to form and grow. They may be pollinated any time during this growth, and will only be fertile, if this pollination takes place. No wonder they survived during those millions of early world years.

Unfortunately the sex of the plants cannot be determined before they produce their pollen cones or, as with the female, their seed cones. Without a knowledge of the varieties even these cones are not obviously recognisable for ascertaining the species' sex. In most species, or varieties, the F cone is the larger. The M cone is taller and slimmer and the F cone broader. In *Cycas,* one variety of the cycads species, the F plant does not grow cones like other varieties. Instead it develops across the centre head of the plant, a fluffy-looking fawn ball, whereas the M forms upright cones in the centre of the plant as do other cycads.

In spring, the new leaf growth appears and continues up from the centre head of the plant, leaving the seeds to appear on short stems circling under the head. Pollination is the same as in all cycads and may take place at any stage before maturity.

With all the species, the F cones continue to form their seed, nuts until matured irrespective of whether pollination has occurred. The cone then opens from the top first, revealing the matured nuts. The flesh or kernel of these is used in cooking and is a reason why they are called "bread palms". It also contains a poison. This has to be removed before the flesh can be consumed. Following the pollination process, the M cone eventually dies and drops off. While there is a shortage of concise printed information available about cycads in most countries of the world, this is not the case in Australia, where there are a number of energetic enthusiasts. Foremost of these is Len Butt, a horticulturist of repute. He has compiled an article entitled "The Cycads of Australia". Len Butt has been a regular contributor to the publication *Australian Plants* which featured his work on Australian cycads. He also writes for many other publications.

His feature article deals specifically with the Australian species of cycads, but they are popular in other parts of the world where they can be found growing alongside local and imported plants. It will be easy to gauge from this how Len Butt's information can be applied to cycads in general.

These hardy plants seem equally at ease in hot deserts, cool and salty ocean fronts, lush glasshouses and sunny or shady positions in home or botanical gardens. I don't know of a hardier plant than the cycad. It would have to be, to have survived unchanged for 170 million years. I hope you enjoy his article.

## THE CYCADS OF AUSTRALIA

Contributed by Len Butt

In Australia the indigenous cycads are scattered throughout the continent, mainly to the north of 30° latitude. They are absent from Victoria and South Australia, but strangely enough, the best known fossil beds of these prehistoric plants are in Victoria.

Cycads are a group of woody stemmed plants of a very ancient flora and related species dating back 200–300 million years, some with thick, columnar caudices or trunks, others with the caudex completely subterranean. The trunks consist of a central core of rather fleshy conducting tissues, on the outside of which is a thick layer of row upon row of persistent leaf bases rising from base to crown. They are mostly unbranched, but multiple heads on mature *Cycas* and *Lepidozamia* are quite common.

Most have the physical appearance of palms, and in the case of *Macrozamia* species such as *M. moorei*, could easily be mistaken for them. However, the *Bowenia* species have rather a fern-like appearance and indeed one of them has long been known as the "Byfield Fern".

The leaves can be called fronds, and in a few species even rise circinately (uncoiling) from the crown as in the true ferns. The beautiful crown of leaves is the primary attraction of these plants. The young fronds are very tender and delicate and susceptible to injury but soon become tough and strong with a glossy surface.

## THE POISONOUS PROPERTIES OF OUR CYCADS

Most of these poisons are contained in the exterior of the seed, and let it be noted indigenous marsupials and other animals that readily eat cycad seed consume the non-toxic flesh of the fruit and leave the actual seed. The hard shiny

*Cycas media*
Australian Nut Palm comes from Queensland.

*Cycas armstrongii*
A female plant with immature seeds. (Photo L. Butt)

*Lepidozamia hopei*
A tall grower (10–20m/30′–60′) from north-eastern Queensland rainforests.

*Macrozamia communis*
A magnificent specimen used in landscaping at a Sydney motel.

fronds of the plants contain little or no poison on maturity, so only at the period of the year when they are producing new centre fronds are they accessible to stock. Even then, this would only refer to the low growing species, as many species with trunk formations are well out of the physical reach of stock.

Unfortunately, it seems, most reports of stock poisoning come from Queensland districts where the taller trunked species grow, the most maligned being *Macrozamia moorei* and *Cycas media*. As the mature leaves are non-toxic, and the new fronds out of reach, the only means of receiving this poison would be through eating new fronds of young seedlings, or completely devouring fallen loose seeds, or eating fresh new tender shoots, from a plant already burnt and blackened by bushfire.

## CYCADS AS FOOD

Although I have noted that Captain Cook's men became ill from eating unprepared seed, this has been a part of the diet of our aboriginal people for probably as long as they have lived in Australia. The starch from seeds and trunk is processed by various methods including boiling, roasting, and continuous soaking and drying out and fermenting. I have read no evidence that the fresh young fronds are ever eaten by native peoples. The large peeled seeds of the Cycad genera in this country are dried in the sun and then soaked in running water for four to five days. The cleaned seed is then pounded into a paste between two rocks and baked under hot ashes to be eaten as cakes. Further information about the *Macrozamia* and *Lepidozamia* is that the process used by other tribes is similar except that the peeled and dried seed is crushed and pounded first, then processed in a "dilly bag" through running water, formed into cakes and eaten after roasting. *Bowenia spectabilis* also was used, but there is no mention as yet that *B. serrulata* was used. This is probably because of its habitats which are only two pockets of country that might not have been frequented by tribes that used it for culinary purposes.

## POLLINATION AND PROPAGATION

Cycads are dioecious. That is to say they consist of separate male and female plants. Males bear narrow pine-like cones which produce sporangia (pollen), and females in all genera except the *Cycas* produce seeds with a larger cone.

Many cycad species have only a single caudex, and therefore the only means of reproduction is by seed. The viability of seed is variable and erratic to say the least. To date there is no evidence that any species of insect is involved in the pollination of cycads.

Male plants generally produce multiple cones while females are either single-coned (as in *Cycas*), or produce relatively few cones in comparison with the males.

At pollination time the male cone elongates and the segments open to deposit large quantities of pollen grains on the fronds of the plant, or to cover the cone, like thick fawn-coloured dust. The period between pollen penetration and fertilisation can run into a few months, and only when the cone breaks up and discharges its fleshy seed can they be termed fertile.

Although many cycads are arid-growing plants, moisture does of necessity play a vital part in their survival. The fleshy fruit that covers all the seed is non-poisonous and attractively coloured, generally orange to orange-red. Nature has left nothing to chance in the survival of these remarkable plants, and relatively few seedlings are found around the parent considering the great quantity of seed expelled at one time. Distribution must be carried out by animals and birds. This, of course, has been proven to be the case in Australia, and, as in Africa where the *Encephalartos* relies on the baboons and thornbill birds, here we have the foraging marsupials and even introduced mammals to help in distribution. Possum and related species carry off the seed, eating the fleshy fruit and discarding the poison-bearing nut kernels. Even the larger marsupials aid in the distribution of the fallen seed.

In cultivation, hand pollination has successfully been carried out by only a few enthusiasts, generally using *Cycas revoluta* as their recipient. Seed from Australian cycads should be given the float test on receipt. That is to say fertile seed will sink if placed in water, but the infertile seed will float because a pocket of space exists between the outer shell and the withering seed core. Much speculation exists regarding the loss of viability of cycad seeds. Local authorities tell us that only fresh seed is viable and must be kept so by storing in damp moss, and indeed, quite a lot of my germinating seed was planted successfully by adhering to this advice.

The removal of seedlings from around a specimen is a quicker method, but one that must be done with diligence so as not to damage the tap root. The rather laborious job of lifting a more mature specimen requires plenty of patience and time and must be carried out with a lot of care so as not to break too much of the tap root. Some cycads have

*Lepidozamia peroffskyana*
Displaying a large F cone.

*Macrozamia paulii-guilielmi* ssp. *paulii-guilielmi*
with F cone (left) and M cone (right)
(Photo Len Butt)

survived when only the top and caudex was secure, but others will develop rot easily if their tap roots are severed.

*Macrozamia lucida*
Pineapple *Zamia* grows in habitats north and south of Brisbane.

*Lepidozamia peroffskyana*
Scaly *Zamia* grows in coastal rainforests of New South Wales and Queensland.

## CYCADS IN OUR LANDSCAPES

Until such time that commonsense conservation of this family of plants becomes the primary function of our government authorities, the best form of preservation we can home for is to use them in our landscape scene, and cultivation at nursery level seems the logical conclusion. Already in Australia there is a stirring of interest for the great palm-like curiosities that are such a tourist eye-catcher on some of our otherwise arid hillsides. And the exclamations of delight that greet us from the uninitiated on seeing their first "Byfield fern" or "Zamia fern" is like ambrosia to the lovers of the cycad family. Their potential as garden plants and landscaping features is enormous. Because of their tenacity for survival their growth is not confined. Local gardeners around Brisbane sometimes have used *Cycas revoluta* as a central frontispiece in their gardens as far back as fifty years ago. On odd occasions I have found large well grown plants of *Macrozamia communis* used, two in particular being a couple of lovely specimens, male and female, side by side on a lawn in Annerley, a suburb of Brisbane's southside. These plants have been there for at least thirty years to my knowledge.

The best place to make the public aware of any genus is to place it in prominent positions in our botanical gardens and parks and even as material for beautification among indigenous plants on freeways. South Australia and New South Wales already have good collections of the cycads, and the old Botanic Gardens in Brisbane have quite a few specimens of such species as *Lepidozamia peroffskyana, L. hopei, Cycas circinalis* and *C. media* and perhaps and odd *Bowenia.*

Around Brisbane the feature palm *Phoenix roebelenii* (Pygmy date) has been used, almost to abandon, between Roma Street Station and King George Square. The beautiful effect this gives makes it very evident that such cycads as *Cycas media, Macrozamia miquelii,* and *Macrozamia moorei* would look equally as outstanding in such situations. One has only to see the sheer dominant beauty of *M. moorei* on the hillslopes of Carnarvon Gorge, or the great stands of *Macrozamia miquelii* on the sandy terrains of Fraser Island, to realize the truth of this.

For any landscape project with an Australian flavour *Cycas* is adaptable, especially now that eucalypts are widely used in and around home garden areas. This genus blends well under the gum trees as also does the *Bowenia* which has a distinct fern-like appearance. Blended with the smaller true ferns such as *Adiantum* (maiden hair), *Pteris* (round fern), and *Davallia* (haresfoot) a very pleasant picture of the Australian bush is achieved. There should be enough tall cover to attain a mottled sunlight effect to really enhance this setting. *Cycas media* is a true coastal species and is now generally accepted as being a slim-trunked plant to about 3 metres high (10'), the female seed appendages being very prominent and visible under the crown of fronds. In coastal situations or on land with a steep slope this cycad could be used successfully as a feature on terraced banks or on top of a mound augmented with rocks. Hardier native plants, not growing higher than the crown of the cycad, could be used as surrounds. Actually, much the same effect is achieved with tree ferns as with cycads, but only a few types of tree fern will stand the sun successfully.

For large properties *Macrozamia moorei* can be used to advantage on *eucalypt*-dotted hill slopes or even as avenue trees up a large and spacious driveway. *M. moorei* has a thick solid trunk and is a very imposing plant. The fronds are erect and arching, the pinnae are blue and V-channelled. On the whole this cycad is probably the most majestic of all the Australian genera. In North Queensland and the more tropical areas there are plenty of landscape types available in the many forms of *Cycas*. The smaller *Cycas armstrongii* which is fairly common in the Arnhem Land and Darwin areas would be ideal as a decorative subject in arid area gardens. One report I have from a Brisbane botanist, is that this plant is deciduous. *Cycas calcicola* and *C. pruinosa* are both species of the far north and north-west, and should serve as excellent landscape material for the tropics. At Gloucester Park, which is a coastal resort-cum-nursery just north of Dingo Beach, Whitsunday Coast, *Cycas normanbyana* grows in the vicinity. A row of large mature specimens is growing just outside the main wooden buildings in the nursery area. These are so picturesque, when I saw them, I at first assumed they were *Cycas cairnsiana,* which is a little better known there.

There is always the situation where a lower plant is needed with the circular symmetry of the smaller grass trees or tree ferns. In these situations the ideal species would be *Macrozamia* mentioned as growing in thickets on the northern New South Wales coast, such as *M. fawcetti,* which grows on coastal areas mainly north and south of Grafton while *M. heteromera,* or *M. stenomera* also occur around Gunnedah and Narrabri to the west. They would suit smaller gardens, in rockeries, possibly among top cover of eucalypts or similar trees.

In Queensland, the species *Macrozamia miquelii* grows

*Macrozamia miquelii*
This cycad has handsome upright fronds growing to 2m (6') in length.

*Macrozamia riedlei*
Zamia Palm from Western Australia in its natural habitat.
(Photo L. Butt)

*Cycas angulata*
Female plant with seeds in natural habitat, Gulf of Carpentaria.
(Photo D. Stallard)

*Cycas angulata*
A group of aged giants near the Gulf of Carpentaria in northern Australia.
(Photo D. Stallard)

in terrains consisting of shaded rock hillsides, dense coastal scrub, with foliage cover high up, moist sandy loam within a hundred metres or yards of fresh-water lakes (as in Fraser Island) and on mountain terrain. The versatility of this plant led people to the belief that there were several species. Such a type could be ideal in moist landscape situations, or even as a patio plant. *M. miquelii* only develops a trunk in its old age, and then only a short one.

When planting cycads in the garden simulate their natural habitat, and growing conditions. Large granite rocks, small boulders, sawn logs with the bark attached, and careful use of ground foliage plants all go to make an ideal natural situation giving atmosphere to your particular plant. There are many of the smaller species ideally suited as potted plants to decorate interior courtyards, verandas, along the front of houses, or as added beauty to the patio situations. So many people living in high rise units still like to have potted plants on their little balconies, and there are even a few types of cycad that could be placed in this position.

First choose the container! This should of necessity have depth to accommodate the long taproot, especially if your plants are seedling-grown. I have always favoured terracotta pots because of their porosity, but these require ample drainage crocks in the bottom, and perhaps a strong piece of gauze over the hole to stop the entry of unwanted pests. The mixture for all of the plants should be friable, and a good example of this is the John Innes base mix. If the cycad is from firm sandy terrain, the addition of more sand in the medium is essential. As the great majority of these plants come from arid to semiarid places, watering should be of a similar level to that given to cacti; plentiful in the hot and windy periods, with a general easing off during winter.

Len Butt has 15 years experience in horticulture, including work in the nursery division of the Queensland Department of Forestry and also in commercial nursery management. He is presently a study group leader on cycads with the Study Group for Australian Plants (SGAP) and has published a book *Cycads of Australia* through the SGAP. He is a prominent member of the Palm and Cycad Society; contributes many articles to the literature and lectures to the members on this subject. He is also conversant on many other horticultural subjects.

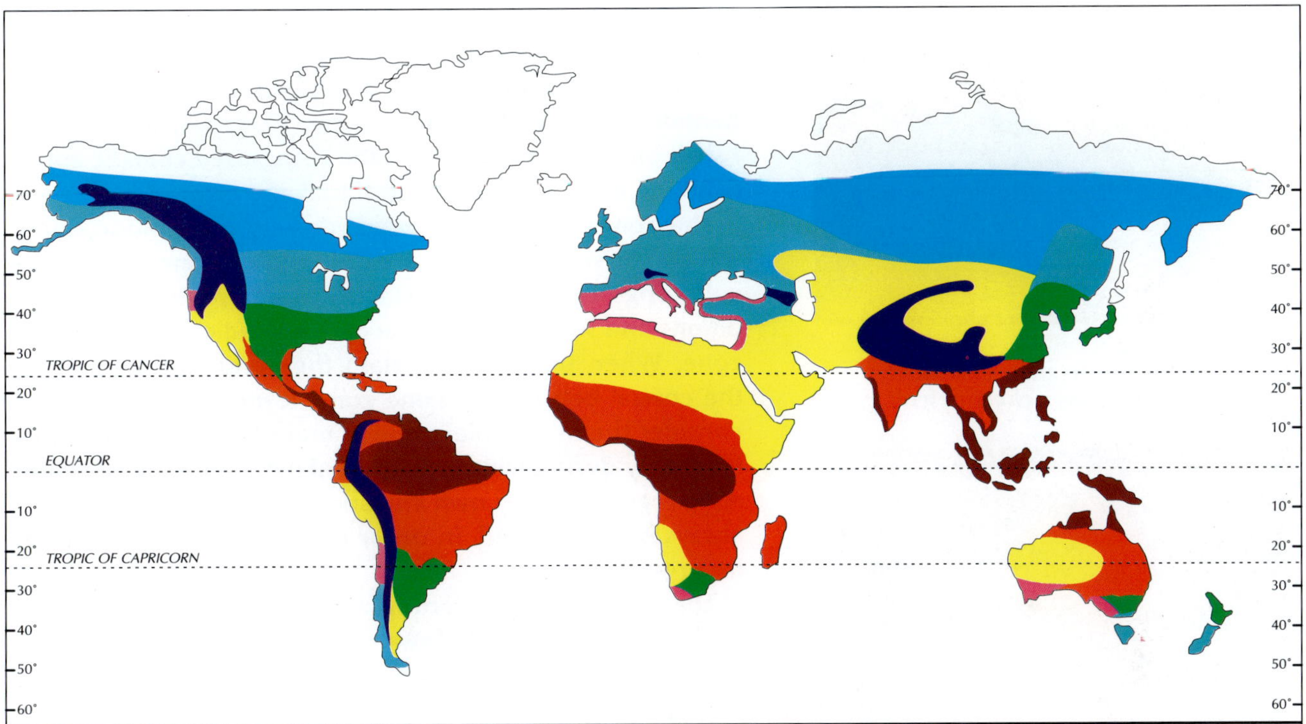


# WORLD CLIMATES

CLIMATIC ZONES

- EQUATORIAL
- TROPICAL incl. Subtropics
- WARM TEMPERATE Mediterranean
- WARM TEMPERATE
- COOL TEMPERATE
- COLD TEMPERATE
- POLAR
- DESERTS OR ARID AREA
- VARIOUS MOUNTAIN CLIMATES

*Cycas calcicola*
Photographed in its natural habitat, this female plant has a circle of fruit. (Photo D. Stallard)

*Cycas calcicola*
In its natural habitat at Tolmer Falls, south-west of Darwin. Note the male cone.
(Photo D. Stallard)

# THE CLIMATIC ZONES

In studying the habitat maps to understand the origins of both palms and cycads, it will become apparent that these species are only naturally located between the latitudes of 40° north and south of the equator. This then indicates that they mainly require a warm or mild climate in which to grow. That is why it is important to know of, and understand, the various climatic zones of the world. These can be very complex if studied in depth as they embrace more than just latitudes and distances from the oceans. Other involved factors are sea current, altitude, prevailing winds, rainy season, humidity, temperature, and so on. However, in this book I shall only touch broadly on the recognised climatic zones, resulting in a certain amount of trial and error.

The zones are colour-coded for easy recognition, but it will also be seen that there are different sections within those colourings. Usually the coastal climates have the milder conditions and would be more accurately indicative of another zone. When trying to ascertain the conditions away from the coastal belt for the same zone, it might be best to also consider the next cooler zone as more typical of that area—especially during winter, when making a planting selection. As every area has its own climatic peculiarities these should always be considered, in order to obtain the best, trouble-free results.

The climatic zones starting with the warmest, or nearest to the equator, are as follows:

- Equatorial or Torrid Zone
- Tropical Zone
- Subtropical area (not really a recognised zone)
- Warm Temperate Zone
- Cool Temperate Zone
- Cold Temperate Zone
- Polar Zone
- Desert areas

**The Equatorial or Torrid Zone** A very humid area around the equator and within the tropics. These areas are usually lowlands, with a humid cloud cover, daily rainfall. They are hot and humid all the time with no definite seasons.

The region includes the cities of Singapore, Penang, Kuala Lumpur, Djakarta, Bali, Port Moresby, Saigon, Panama, Caracas, Luanda, Nairobi, Addis Ababa, Dar es Salaam, Lima, Madras and Colombo. The highlands in this region are a little cooler and less humid and fall into the tropical zone, or cooler, according to altitude.

**The Tropics** Includes the less torrid areas within the tropics 24° north and south. The seasonal day temperatures are hot and humid but a little cooler at night and during the short winter period. Temperatures could be higher inland, or in desert areas with cooler nights. The area is monsoonal and the rain seasonal. Cities within the region include: Guangzhou, Hong Kong, Manila, Taipei, Darwin, Cairns, Townsville, Rockhampton, Suva, Honolulu, Acapulco, Havana, Rio de Janiero, Dakar, Khartoum, Aden, Salisbury, Bombay, Calcutta, Rangoon and Hanoi. The highlands' climate is usually in a cooler zone, according to altitude.

**The Subtropics** A fringe area of the tropical and warm temperate zones, where many of the tropical plants may be grown in the garden, providing they are not subjected to cooler winter temperatures and winds. It is an idyllic living climate, with pleasant summer temperatures and mild short winters, and regular rainfalls. Cities within the region include: Brisbane, Coffs Harbour, Geraldton, New Orleans, Miami, Cairo, Durban, Bahrain, Karachi and Foochow. Altitudes and inland climates with extremes may fall into a cooler zone. Latitudes 24° to 30° north and south of the equator.

**Warm Temperate Zone** The extremes begin to become more noticeable in the Temperate Zones, with milder coastal temperatures of the four seasons is about even (fanned by cool winds) than those inland. The temperature in summer is hot and the winter, cool. There is seldom snow in the winter, but some frosts could be expected. Cooler or elevated areas would favour a cooler zone. Cities within the region are: Sydney, Perth, Melbourne, Cape Town, Auckland, Seoul, Tokyo, Shanghai, Beijing, Los Angeles, Washington, San Francisco, Buenos Aires, Santiago, Lisbon, Athens, Alexandria, Tehran. Latitudes between 30° and 40° north and south, of the equator.

**Cool Temperate Zone** This region is too cool for all but a few of the palm and cycad species in outdoor plantings. Care in selection, plus a little trial and error and the selection of warm, sheltered and even suntrap positions may be necessary. This zone has very large temperature variations within it. The four seasons are well defined.

Planting recommendations would be best sought from local nursery workers, or advisers. Plantings in greenhouses or pots will enhance the safe range and provide indoor plants, which may be summer-spelled outdoors. Frosts and light snow are normal and can be heavier in cooler sections. Cities within the zone are: Hobart, Christchurch, Wellington, London, Paris, Madrid, Vienna, Rome, New York, Montreal, Chicago, Vancouver and Vladivostok. Latitudes 40° to 50° north and south of the equator.

*Trachycarpus wagnerianus*
Will withstand cold and frost

*Macrozamia paulii-guilielmi* ssp. *plurinervia*
A small plant for shaded positions.

*Hyophorbe lagenicaulis*
Bottle Palm is a most unusual palm.

*Encephalartos munchii*
A plant from Mozambique.

**Cold Temperate Zone** This zone would be too cold for outdoor plantings of palms and cycads with only a few exceptions. Warmed greenhouses and indoor plantings would be needed with cross reference of the origins of each species needed. The potted specimens may be spelled outdoors when the temperatures are warm enough, avoiding sun scorch, and returned to warmer shelter areas before cold sets in. Many regions receive prolonged snow. The summers are short and temperatures very extreme. Cities within the zone are: Edmonton, Winnipeg, Glasgow, Copenhagen, Warsaw, Moscow, Berlin and Belfast. Latitudes 50° to 60° N & S of the equator.

**Desert Regions** While this region falls within the above zones, it has its own characteristics. The day and night temperatures are harsh and extreme, but some cycad and palm species originate from these environments. Select typical species for successful plantings if regular water supplies are available. Many desert areas have been turned into arable areas with chemicals and water. Israel is an example of this. Deserts are arid, sandy or vegetation-free areas, or semiarid with only scrub vegetation and the occasional tree. There are rivers, or seasonal water courses, waterholes, or just subterranean water. Cities within desert regions include Alice Springs, Las Vegas, Tucson, Tel Aviv, Arab cities and oil towns, many too small and little-known to mention them all.

**Assess your Climatic Zone** Firstly, link your locality climatically into one of the above zones and regions on the map. Decide whether you have a mild climate, like the coast offers, or an inland climate, or a cooler elevated region. After deciding your zone and region, then look up the Quick Reference Guide of palms and cycads for that region. In the case of cycads, concise information for all species is not always available owing to the inability to obtain a full selection for study. As time goes by this information will be added to, for both cycads and palms.

Lastly, check the particular variety in the text, to find out its origin, and help you to decide whether it may be a suitable variety for the use you have in mind. Perhaps, if it is not suitable for an open position, a sheltered suntrap may be better. Even a sheltered veranda or indoor position could be the answer. It is usually possible to grow a palm quite successfully in the next zone, from that of its natural habitat. Varieties even further afield than this have proved suitable in certain instances with some palms and cycads appearing to have a greater hardiness to colder regions, not being set back by frost or snow. The plants may be kept in pots for a trial period before planting them permanently in the garden.

# CHOOSING PLANTS

## QUICK REFERENCE GUIDE

This quick reference guide has been added as a speedy indication of the climatic zones, types, sizes, planting needs for outdoors or indoors, and other uses for that particular palm or cycad. It is not possible to indicate the specific information to each and every palm in the book, this type of research has never been carried out. You will still need to refer to the origin for any other information to make a judgement as to whether it would be successful for your purpose. The information supplied is known information and a blank will be left if the palm is unsuitable for that category, or the information is unavailable. On some occasions more than one category will apply to that type of palm, because of variety differences. In many cases an "x" will be used to designate that this category applies to a lesser number of the species than the . marking. It must be remembered that generalisations must apply with the genera mentioned.

**Seaside planting tolerance** — 1–3 the lower the number, the greater the tolerance of that palm, or certain of its species, to salt and wind.

**The coloured boxes** indicate the natural habitat where most of those species are to be located. This should make selection for planting conditions much more simple. In many cases more than one box is marked, indicating that different varieties of the species may be located over a wide range of climatic zones. Refer to the text for these species.

**Armed with spikes,** will give a warning to keep that particular palm or cycad away from traffic areas thereby avoiding injuries to family and friend.

***Frost tolerant,** indicating that this species will not be permanently injured by frosts. Refer to the text, as some may also withstand some snow.

**Dry-area tolerant,** indicates that this species originates from and will tolerate, once established, drier conditions. This does not mean not to water the plant regularly in cultivation, but it should be planted in an area that drains freely. Water should be applied deeply during droughts.

**Wet-area tolerant,** indicates that it may be planted in lower-lying occasionally soggy areas. It requires proper drainage methods when planting, not soggy holes, which don't drain. Only in rare instance does it mean permanently wet swamps.

**Hot-sun tolerant,** indicates that the plant can tolerate direct sun from an early age.

*Latania verschaffeltii*
A showy Yellow Latan Palm, so named because the juvenile growth is yellow.

*Zamia* hybrid species
Good in a rockery aspect. A seed cross.

| PALM VARIETY | EQUATORIAL ZONE | TROPICAL ZONE | SUBTROPICAL AREA | WARM TEMPERATE | COOL TEMPERATE | COLD TEMPERATE | SINGLE TRUNK | CLUSTERED TRUNKS | LARGE GROWTH | MODERATE GROWTH | SMALL GROWTH | CLIMBING GROWTH | SEASIDE TOLERANT | DRY AREA TOLERANT | WET AREA TOLERANT | HOT SUN TOLERANT | FULL SHADE TOLERANT | ARMED WITH SPIKES | FEATHER LEAVES | FAN LEAVES | OTHER SHAPE LEAF | INDOOR PROVEN |
|---|---|---|---|---|---|---|---|---|---|---|---|---|---|---|---|---|---|---|---|---|---|---|
| ACOELORRAPHE | ● | ■ ● | ● | ● | | | | ● | | ● | | | 2 | | ● | ● | | ● | | ● | | |
| ACROCOMIA | ● | ■ ● | ● | ● | | | ● | | ● | ● | | | 2 | | | | | ● | ● | | | |
| ACTINOKENTIA | ● | ■ ● | ● | ● | | | ● | | | | ● | | | | | | | | ● | | | |
| ACTINORHYTIS | ■ ● | ● | ● | ● | | | ● | | ● | | | | | | | | | | ● | | | |
| AIPHANES | ● | ■ ● | ● | ● | | | ● | | | ● | ● | | | | | | | ● | ● | | | |
| ALLAGOPTERA | | ■ ● | ● | ● | | | | ● | | | | | 1 | | | ● | | | ● | | | |
| ARCHONTOPHOENIX | ● | ■ ● | ■ ● | ● | | | ● | | ● | ● | | | | | ● | | ● | | ● | | | |
| ARECA | ■ ● | ● | ● | ● | | | ✖ | ✖ | | ● | ● | | | | | | | | ● | | | |
| ARENGA | ● | ■ ● | ● | ● | | | ✖ | ✖ | ● | ● | ● | | | | | ● | | | ● | | | ● |
| ASTROCARYUM | ■ ● | ● | ● | | | | ● | | | ● | | | | | | | | ● | ● | | | |
| ATTALEA | ■ ✖ | ■ ✖ | ● | | | | ● | | ✖ | ✖ | | | | | | | | | ● | | | |
| BACTRIS | ■ ● | ● | | | | | ✖ | ✖ | ● | ● | ● | | | | | | | ● | ● | | | |
| BALAKA | ● | ■ ● | ● | ● | | | ● | | | ● | ● | | | | | | | | ● | | | |
| BASSELINIA | | ■ ● | ● | ● | | | ✖ | ✖ | | ● | ● | | | | | | | | ● | | | |
| BENTINCKIA | ■ ● | ● | ● | | | | ● | | | ● | | | | | | | | | ● | | | |
| BISMARCKIA | ● | ■ ● | ● | ● | | | ● | | ● | | | | | | | | | | | ● | | |
| BORASSUS | ■ ● | ● | | | | | ● | | ● | | | | 2 | | | ● | | | | ● | | |
| BRAHEA ✱ | ● | ● | ● | ■ ● | ● | | ● | | | ● | | | 2 | ● | | ● | | ● | | ● | | |
| BURRETIOKENTIA | ● | ■ ● | ● | | | | ● | | ● | ● | | | | | | | | | ● | | | |
| BUTIA ✱ | ● | ■ ● | ● | ● | ● | | ● | | ● | ● | | | 3 | ● | | ● | | ✖ | ● | | | |
| CALAMUS | ■ ● | ● | ● | ● | | | ✖ | ✖ | | | ● | | | | | | ● | ● | | | | ● |
| CALYPTROCALYX | ■ ● | ● | | | | | ✖ | ✖ | | ● | ● | | | | | | ● | | ✖ | | | |
| CALYPTROGYNE | ● | ■ ● | ● | ● | | | ● | | | | ● | | | | | | ● | | ● | | | |
| CALYPTRONOMA | ● | ■ ● | ● | | | | ● | | | ● | | | | | | | | | ● | | | |
| CARPENTARIA | ■ ● | ● | ● | ● | | | ● | | | ● | | | | | | | | | ● | | | |
| CARYOTA | ■ ● | ● | ● | ● | | | ✖ | ✖ | ● | ● | | | | | | ✖ | ✖ | | | | ● | ● |
| CEROXYLON | ■ ● | ● | ● | ● | | | ● | | ● | | | | | | | | | | ● | | | |
| CHAMAEDOREA | ● | ■ ● | ● | ● | | | ✖ | ✖ | | | ● | ✖ | | | | | ● | | ● | | ✖ | ● |
| CHAMAEROPS ✱ | | ● | ● | ■ ● | ● | | | ● | | ● | | | 2 | ● | | ● | | ● | | ● | | |
| CHAMBEYRONIA | ● | ■ ● | ● | ● | | | ● | | | ● | | | | | | | | | ● | | | |
| CHELYOCARPUS | ■ ● | ● | ● | | | | ✖ | ✖ | | ● | | | | | | | | | | ● | | |
| CHRYSALIDOCARPUS | ● | ■ ● | ● | ● | | | ● | ● | ● | ● | ● | | 3 | | | | ● | | ● | | | ● |
| CHUNIOPHOENIX | | ■ ✖ | ✖ | | | | | ● | | | ● | | | | | | ● | | | ● | | |

■ – natural habitat
● – suitable in most cases
✖ – suitable in less cases
✱ – tolerates some frost
no number – unsuitable or doubtful
1 – very tolerant
2 – less tolerant
3 – slightly tolerant

**PALM VARIETY**

■ – natural habitat
● – suitable in most cases
✖ – suitable in less cases
✱ – tolerates some frost
no number – unsuitable or doubtful
1 – very tolerant
2 – less tolerant
3 – slightly tolerant

| PALM VARIETY | EQUATORIAL ZONE | TROPICAL ZONE | SUBTROPICAL AREA | WARM TEMPERATE | COOL TEMPERATE | COLD TEMPERATE | SINGLE TRUNK | CLUSTERED TRUNKS | LARGE GROWTH | MODERATE GROWTH | SMALL GROWTH | CLIMBING GROWTH | SEASIDE TOLERANT | DRY AREA TOLERANT | WET AREA TOLERANT | HOT SUN TOLERANT | FULL SHADE TOLERANT | ARMED WITH SPIKES | FEATHER LEAVES | FAN LEAVES | OTHER SHAPE LEAF | INDOOR PROVEN |
|---|---|---|---|---|---|---|---|---|---|---|---|---|---|---|---|---|---|---|---|---|---|---|
| CLINOSTIGMA | ● | ■ ● | ● | | | | ● | | | ● | | | | | | | | | ● | | | |
| COCCOTHRINAX | ● | ■ ● | ● | ● | | | ✖ | ✖ | | ● | ● | | 1 | | | | | | | ● | | |
| COCOS | ■ ● | ● | ● | | | | ● | | | ● | | | 1 | | | ● | | | ● | | | |
| COPERNICIA | ● | ■ ● | ● | ● | | | ● | | ● | ● | | | 2 | ● | | ● | | ● | | ● | | |
| CORYPHA | ■ ● | ● | ● | | | | ● | | ● | | | | | ● | | ● | | ● | | ● | | |
| CRYOSOPHILA | ■ ● | ● | ● | | | | ● | | ● | | | | | | | | | ● | | ● | | |
| CYPHOPHOENIX | ● | ■ ● | ● | | | | ● | | | ● | | | | | | | | | ● | | | |
| CYPHOSPERMA | ● | ■ ● | ● | | | | ● | | | ● | | | | | | | | | ● | | | |
| CYRTOSTACHYS | ■ ● | ● | | | | | ✖ | ✖ | ● | ● | | | | | ● | | | | ● | | | |
| DAEMONOROPS | ■ ● | ● | | | | | ✖ | ✖ | | | | ● | | | | | | ● | ● | | | |
| DICTYOSPERMA | ● | ■ ● | ● | ● | | | ● | | | ● | | | 1 | | | | | | ● | | | ● |
| DYPSIS | ● | ■ ● | ● | | | | ✖ | ✖ | | ● | ● | | | | | | ● | | ● | | ✖ | |
| ELAEIS | ■ ● | ● | ● | | | | ● | | ● | ● | | | 2 | | | ● | | ● | ● | | | |
| EUTERPE | ● | ■ ● | ● | | | | ✖ | ✖ | ● | ● | | | | | | | | | ● | | | |
| GASTROCOCOS | ● | ■ ● | ● | | | | ● | | ● | | | | | | | | | ● | ● | | | |
| GAUSSIA | ● | ■ ● | ● | | | | ● | | ● | ● | | | | | | | | | ● | | | |
| GEONOMA | ● | ■ ● | ● | ● | | | ✖ | ✖ | | ● | ● | | | | | | ● | | ● | | ✖ | |
| GRONOPHYLLUM | ■ ● | ● | ● | ● | | | ✖ | ✖ | ● | ● | ● | | | | | ● | ● | | ● | | | |
| GULUBIA | ■ ● | ● | | | | | ● | | | | | | | | | | | | ● | | | |
| HEDYSCEPE | ● | ■ ● | ● | ● | | | ● | | | ● | | | 2 | ● | | | | | ● | | | ● |
| HETEROSPATHE | ■ ● | ● | ● | | | | ● | | | ● | ● | | | | | | ● | | ● | | | |
| HOWEA | | ■ ● | ● | ● | | | ● | | | ● | | | 2 | | | | ● | | ● | | | ● |
| HYDRIASTELE | ■ ● | ● | ● | | | | | ● | | ● | ● | | | | ● | | ● | | ● | | | |
| HYOPHORBE | ● | ■ ● | ● | ● | | | ● | | | ● | | | 1 | | | ● | | | ● | | | |
| HYPHAENE | ● | ■ ● | ● | ● | | | ✖ | ✖ | ● | ● | ● | | 2 | ● | | ● | | ● | | ● | | |
| IRIARTEA | ■ ✖ | ■ ✖ | | | | | ● | | ● | | | | | | ● | | | | ● | | | |
| JOHANNESTEIJSMANNIA | ■ ● | ● | | | | | ● | | | ● | | | | | | | ● | ● | | | ● | |
| JUBAEA ✱ | | ● | ● | ■ ● | | | ● | | ● | | | | | ● | | ● | | | ● | | | |
| JUBAEOPSIS | | ■ ● | ● | | | | | ● | | ● | | | | | | | | | ● | | | |
| KENTIOPSIS | ● | ■ ● | ● | ● | | | ● | | ● | | | | | | | | | | ● | | | |
| KORTHALSIA | ■ ● | ● | | | | | | ● | | ● | ● | ● | | | | | | ● | ● | | | |
| LACCOSPADIX | | ■ ● | ● | ● | | | | ● | | ● | ● | | | | | | ● | | ● | | | ● |
| LATANIA | ● | ■ ● | ● | ● | | | ● | | | ● | | | 2 | | | ● | | | | ● | | |

| PALM VARIETY | EQUATORIAL ZONE | TROPICAL ZONE | SUBTROPICAL AREA | WARM TEMPERATE | COOL TEMPERATE | COLD TEMPERATE | SINGLE TRUNK | CLUSTERED TRUNKS | LARGE GROWTH | MODERATE GROWTH | SMALL GROWTH | CLIMBING GROWTH | SEASIDE TOLERANT | DRY AREA TOLERANT | WET AREA TOLERANT | HOT SUN TOLERANT | FULL SHADE TOLERANT | ARMED WITH SPIKES | FEATHER LEAVES | FAN LEAVES | OTHER SHAPE LEAF | INDOOR PROVEN |
|---|---|---|---|---|---|---|---|---|---|---|---|---|---|---|---|---|---|---|---|---|---|---|
| LEPIDORRHACHIS | | | | ■ ● | | | ● | | | ● | | | | | | | | | ● | | | |
| LICUALA | ■ ● | ● | ● | ✖ | | | ✖ | ✖ | | ● | ● | | | | ● | | ● | ✖ | | ● | | ● |
| LINOSPADIX | ● | ■ ● | ● | ● | | | ✖ | ✖ | | | ● | | | | | | ● | | ● | | ● | ● |
| LIVISTONA ✖ ✱ | ■ ● | ■ ● | ■ ● | ● | ✖ | | ● | | ● | ● | | | 2 | ✖ | ✖ | ✖ | | ✖ | | ● | | |
| LODOICEA | ■ ● | ● | | | | | ● | | ● | | | | | | | | | | | ● | | |
| LYTOCARYUM | | ■ ● | ● | ● | | | ● | | | | ● | | | | | | ● | | ● | | | ● |
| MAURITIELLA | ■ ● | ● | | | | | | ● | | ● | | | | | | | | ● | | ● | | |
| METROXYLON | ■ ● | ● | ● | | | | ✖ | ✖ | ● | | | | | | ● | ● | | ● | ● | | | |
| NANNORRHOPS ✱ | | ■ ✖ | ✖ | | | | | ● | | ● | | | | ● | | ● | | | | ● | | |
| NEODYPSIS | | ■ ● | ● | ● | | | ✖ | ✖ | ● | ● | ● | | | | | | | | ● | | | |
| NEONICHOLSONIA | ■ ● | ● | ● | | | | ● | | | | ● | | | | | | | | ● | | | |
| NEOPHLOGA | | ■ ● | | | | | ✖ | ✖ | | | ● | | | | | | ● | | ● | | | |
| NEOVEITCHIA | | ■ ● | | | | | ● | | | ● | | | | | | | | | ● | | | |
| NEPHROSPERMA | ■ ● | ● | ● | | | | ● | | | ● | | | | | | | | ● | ● | | | |
| NORMANBYA | ■ ● | ● | ● | ● | | | ● | | | ● | | | | | ● | | ● | | ● | | | |
| NYPA | ■ ● | ● | ● | | | | | ● | ● | | | | 2 | | ● | | | | ● | | | |
| OENOCARPUS | ■ ● | ● | | | | | ✖ | ✖ | ● | ● | | | | | | | | | ● | | | |
| ONCOSPERMA | ■ ● | ● | ● | | | | | ● | ● | | | | | | ● | ● | | ● | ● | | | |
| ORANIA | ■ ● | ● | | | | | ● | | ● | ● | ● | | 3 | | | | ✖ | | ● | | | |
| ORANIOPSIS | ■ ● | ● | ● | | | | ● | | | ● | | | | | ● | | | | ● | | | |
| ORBIGNYA | ■ ● | ● | ● | | | | ● | | ● | ● | | | | | ● | | | | ● | | | |
| PELAGODOXA | ■ ● | ● | ● | | | | ● | | | ● | | | | | | | | | | | ● | |
| PHOENICOPHORIUM | ■ ● | ● | ● | | | | ● | | | ● | | | | | | | | ● | | | ● | |
| PHOENIX ✖ ✱ | ■ ● | ■ ● | ● | ● | ● | | ✖ | ✖ | ● | ● | ✖ | | 2 | ● | ✖ | ● | ✖ | ● | ● | | | ● |
| PHOLIDOCARPUS | ■ ● | ● | | | | | ● | | ● | | | | | | ● | | | ● | | ● | | |
| PIGAFETTA | ■ ● | ● | | | | | ● | | ● | | | | | | | | | ● | ● | | | |
| PINANGA | ■ ● | ● | ● | ✖ | | | ✖ | ✖ | | ● | ● | | | | | | ● | | ● | | ● | ● |
| POLYANDROCOCOS | ■ | ● | ● | | | | ● | | ● | ● | | | | | | | | ✖ | ● | | | |
| PRESTOEA | ■ ● | ● | ● | | | | ✖ | ✖ | | ● | ● | | | | | | | | ● | | | |
| PRITCHARDIA | ● | ■ ● | ● | ● | | | ● | | | ● | | | 2 | | | ● | | | | ● | | |
| PSEUDOPHOENIX | | ■ ● | ● | | | | ● | | | ● | | | 2 | ● | | | | | ● | | | |
| PTYCHOCOCCOS | ■ ● | ● | ● | | | | ● | | | ● | | | | | | | | | ● | | | |
| PTYCHOSPERMA | ■ ● | ● | ● | ● | | | ✖ | ✖ | | ● | ● | | | | ✖ | | ● | | ● | | | ● |

■ – natural habitat
● – suitable in most cases
✖ – suitable in less cases
✱ – tolerates some frost
no number – unsuitable or doubtful
1 – very tolerant
2 – less tolerant
3 – slightly tolerant

**PALM VARIETY**

■ – natural habitat
● – suitable in most cases
✖ – suitable in less cases
✻ – tolerates some frost
no number – unsuitable or doubtful
1 – very tolerant
2 – less tolerant
3 – slightly tolerant

| PALM VARIETY | EQUATORIAL ZONE | TROPICAL ZONE | SUBTROPICAL AREA | WARM TEMPERATE | COOL TEMPERATE | COLD TEMPERATE | SINGLE TRUNK | CLUSTERED TRUNKS | LARGE GROWTH | MODERATE GROWTH | SMALL GROWTH | CLIMBING GROWTH | SEASIDE TOLERANT | DRY AREA TOLERANT | WET AREA TOLERANT | HOT SUN TOLERANT | FULL SHADE TOLERANT | ARMED WITH SPIKES | FEATHER LEAVES | FAN LEAVES | OTHER SHAPE LEAF | INDOOR PROVEN |
|---|---|---|---|---|---|---|---|---|---|---|---|---|---|---|---|---|---|---|---|---|---|---|
| RAPHIA | ■● | ● | | | | | ✖ | ✖ | ● | | | | | | ● | | | | ● | | | |
| RAVENEA | | ■● | ● | | | | ● | | ● | ● | | | | | | | | | ● | | | |
| REINHARDTIA | ● | ● | ● | ● | | | ✖ | ✖ | | ● | ● | | | | | | ● | | | | ● | ● |
| RHAPIDOPHYLLUM | | ● | ● | | | | | ● | | ● | | | | | ● | | | ● | | ● | | |
| RHAPIS | ● | ■● | ● | ● | | | | ● | | | ● | | 2 | ● | | | ● | | | ● | | ● |
| RHOPALOSTYLIS ✻ | | ● | ● | ■● | ● | | ● | | ● | ● | | | 1 | | ● | | | | ● | | | |
| ROSCHERIA | ■● | ● | ● | | | | ● | | | ● | ● | | | | | | | ● | ● | | | |
| ROYSTONEA | ● | ■● | ● | ● | | | ● | | ● | | | | 2 | | ● | ● | | | ● | | | |
| SABAL ✻ | ● | ■● | ■● | ■● | ● | | ● | | ● | ● | | | 1 | ✖ | ✖ | ● | | | | ● | | ● |
| SALACCA | ■● | ● | ● | | | | | ● | | ● | ● | | | | ● | | ● | ● | | | ● | |
| SATAKENTIA | | ■● | ● | ● | | | ● | | | ● | | | | | | | | | ● | | | |
| SCHEELEA | ■● | ■● | | | | | ● | | ● | | | | | ✖ | | | | | ● | | | |
| SCHIPPIA | | ■● | | | | | ● | | | ● | | | | | | | ● | | | ● | | |
| SERENOA | ● | ■● | ● | ● | | | | ● | | ● | | | 1 | | ● | ● | | ● | | ● | | |
| SIPHOKENTIA | ■● | ● | ● | ● | | | ● | | | ● | | | | | | | | | ● | | | |
| SOCRATEA | ■● | ● | | | | | ● | | | ● | | | | | ● | | | ● | ● | | | |
| SYAGRUS | ● | ■● | ● | ● | | | ● | ✖ | ● | ● | ● | | 3 | | | ● | ● | ✖ | ● | | | ● |
| THRINAX | ● | ■● | ● | ● | | | ● | | | ● | ● | | 1 | ● | | | | ✖ | | ● | | |
| TRACHYCARPUS ✻ | | | ■● | ● | ■● | ● | ✖ | ✖ | | ● | ● | | 2 | | | ● | | | | ● | | |
| TRITHRINAX | | ■● | ● | ● | | | ● | | | ● | | | | | | ● | | ● | | ● | | |
| VEITCHIA | ■● | ● | ● | ● | | | ● | | | ● | | | 2 | | | | | | ● | | | |
| VERSCHAFFELTIA | ■● | ● | | | | | ● | | | ● | | | | | | | | ● | ● | | | |
| WALLICHIA | ■● | ● | ● | ● | | | ✖ | ✖ | | ● | ● | | | | | | ● | | ● | | | |
| WASHINGTONIA ✻ | ● | ■● | ■● | ● | ● | | ● | | ● | | | | 1 | ● | | ● | | ● | | ● | | ● |
| WETTINIA | ■● | ● | | | | | ✖ | ✖ | ● | ● | | | | | ● | | | | ● | | | |
| WODYETIA | ■● | ■● | ● | | | | ● | | | ● | | | | ● | | | | | ● | | | |
| ZOMBIA | | ■● | ● | | | | | ● | | ● | | | | ● | | | | ● | | ● | | |

**CYCAD VARIETY**

■ – natural habitat
● – suitable in most cases
✖ – suitable in less cases
✱ – tolerates some frost
no number – unsuitable or doubtful
1 very tolerant
2 – less tolerant
3 – slightly tolerant

| CYCAD VARIETY | EQUATORIAL ZONE | TROPICAL ZONE | SUBTROPICAL AREA | WARM TEMPERATE | COOL TEMPERATE | COLD TEMPERATE | SINGLE TRUNK | CLUSTERED TRUNKS | LARGE GROWTH | MODERATE GROWTH | SMALL GROWTH | CLIMBING GROWTH | SEASIDE TOLERANT | DRY AREA TOLERANT | WET AREA TOLERANT | HOT SUN TOLERANT | FULL SHADE TOLERANT | ARMED WITH SPIKES | FEATHER LEAVES | FAN LEAVES | OTHER SHAPE LEAF |
|---|---|---|---|---|---|---|---|---|---|---|---|---|---|---|---|---|---|---|---|---|---|
| BOWENIA | | ■● | ● | ● | | | | ● | | | ● | | | | | | ● | | ● | | |
| CERATOZAMIA | ■● | ■● | ● | | | | ● | | | ● | | | | | | | ● | ● | ● | | |
| CYCAS | ■● | ■● | ■● | ● | | | ✖ | ✖ | | ✖ | ✖ | | 3 | ✖ | | ✖ | ✖ | ● | ● | | |
| DIOON | ■● | ■● | ● | ● | | | ✖ | ✖ | | ✖ | ✖ | | | ✖ | | ✖ | ✖ | ● | ● | | |
| ENCEPHALARTOS ✖ ✱ | ■● | ■● | ■● | ■● | | | ✖ | ✖ | | ✖ | ✖ | | ✖ | ✖ | | ✖ | ✖ | ● | ● | | |
| LEPIDOZAMIA | | ■● | ■● | ● | | | ● | | | ✖ | ✖ | | | | | | ✖ | | ● | | |
| MACROZAMIA | | ■● | ■● | ■● | | | ● | | | ✖ | ✖ | | | ✖ | | ✖ | ✖ | | ● | | |
| MICROCYCAS | | ■● | ● | ● | | | ● | | | | ● | | | | | | ● | ● | ● | | |
| STANGERIA | | ● | ● | ■● | | | ● | | | | ● | | | | | | ● | | ● | | |
| ZAMIA | ■● | ■● | ■● | ● | | | ● | | | ✖ | ● | | 1 | ✖ | | ✖ | ● | ✖ | ● | | |

# LANDSCAPING WITH PALMS AND CYCADS

Palms and cycads are not related although they may look similar in many respects. I have heard comments that cycads are "bonsai palms". As they complement each other, they are wonderful to use together in landscape ventures. Most of the cycads, along with many other shorter plants and shrubs, benefit from the intermittent shade from palms, which also have a compact root system. It is interesting to wander through the lovely botanical gardens of our leading cities and see the way palms and cycads are used in landscape. On a recent visit to Fairchild Gardens in Miami, Florida, I found it intriguing to see the impact *Encephalartos ferox* had on the passing public, who were stopped, spellbound by the large bright orange-red seed cones. These were just opening out, with the seeds spilling forth. The lovely *Lepidozamias* of Australia have a larger and just as fascinating seed cone only this is green. It is more rewarding to attempt to grow a married pair of cycads in order to enjoy the beauty and pleasure from these seed cones, as well the satisfaction of being able to present a seed to a friend who has admired the specimen.

Just as interesting are some of the palm species. Luckily, many of these are self pollinating, only needing one plant to produce seeds. However, this is not the case with all species. All this pollination information is available in the palm text for all species.

I'm sure everyone would love to have their own home-grown coconuts. This is not possible of course unless you have a tropical climate, to suit coconuts. That is why you need to consult the climatic zones before choosing palms to plant in landscape aspects

In the following chapters I will explain further about landscaping, with both palms and cycads plantings, how to create your own rainforest, pot growing, potting mixes, indoor use, pests and trouble-shooting, as well as providing that quick reference chart for zonal recommendations. As full information on cycads from all areas of the world is very sparse, and in need of greater research, I have included an article on Cycads in Australia by a leading authority, Len Butt. You should find this very interesting and informative. While it does not deal with varieties from other countries, the subject dealt with, will apply to most of the cycads known. Their treatment is much the same, wherever they came from.

In these articles, I will specifically deal with palms and cycads, not all plants for general landscaping. However, the

*Chambeyronia macrocarpa*
The new leaf is bright red.

*Chrysalidocarpus lucubensis*
A lovely palm for the tropics or subtropics.

*Encephalartos princeps*
A showy plant in the right setting.

*Archontophoenix alexandrae*
Note the two colourful bunches of almost ripe fruits.

planting instructions will be relevant for all types of plants. Mix these two species in a general landscape plant, or vice versa and once your palm and cycad plan is complete then come back and add your other plants to best effect. That should add colour. Most of the planting information given however will apply generally, for successful results with all plants, in your planting techniques.

## SELECTING A POSITION

Let us begin our landscaping with the assumption that you have already drawn up an overall plan, which includes some palms or cycads, along with other plants. Or, perhaps you have an already established garden and can't resist adding a palm, or cycad, or two. This is called spot-planting. Maybe you have a shady section where the lawn grasses won't grow properly, or one of those suntrap areas sheltered from cold winds. Perhaps you are fed up with mowing the lawn? Why not turn it into your own rainforest under a canopy of palms. This is a wonderful project for the person who just can't resist buying a favoured plant, whenever they see one. Additional palms can usually be fitted into such an area. Even the discarded indoor plant has a chance of survival in that kind of garden setting, if the climate is right.

When starting a garden from scratch and wishing to plant a few palms, or cycads, refer to any of the quick reference sections of this book that may be indicative of your needs, climatically, because of drainage problems, or questions of a similar nature. After you have made your selection, supply and preparation come next. If you reside in the cooler areas, supplies may have to be ordered, but don't let this deter you. It is far better to have the palm of your choice, than to settle for any old palm that the nursery happened to have in stock at the time. They will soon get it for you, even if they have to bring it from areas where palms are an every day selling line.

I don't believe that some regions have felt the full impact of having lovely palms included in their gardens. In the past many areas have been restricted to a limited range of palm species, within a general range of native plants, or popular shrubs and trees.

Now palms are not just looked upon as being one or two species, such as one large (probably a date or fan palm), and a smaller palm (perhaps a dwarf date or some small local native species). To the earlier old fashioned nursery-worker, a palm of any kind was just a "palm", in the selling range and preferably the easiest one to get seed supplies of at that time. Today we are all more choosy and expect our suppliers to cater for our selections, and rightly so. How else do we raise our standards, or the suppliers their standards. They will get your choice for you, or another will.

I don't imply that you could obtain every palm or cycad mentioned in this book. Many foreign varieties won't be seen, as yet, other than in lovely botanical gardens or private collections. You can however, make your selection from a much larger range than the old-fashioned nursery used to stock. Different countries will carry different selections, but the exotic varieties are on the "up" everywhere. Every region appears to have its own selection of "lovelies" you can pick from. All the varieties shown are being grown somewhere, as part of the general range available in that area.

When selecting a position for plantings, remember palms can grow to various heights, some very tall, whether single-trunked, or multiple-trunked. They can also come as lower and thicker clumps, or very small 1m (3′) sizes. This is where the visual sighting of the palms and also cycads pictured in this book is of the utmost importance, to help you to select the palm to suit the position, whether it be in shade, full sun or under a canopy of trees. The lists can further assist for cross referencing. Also, don't forget the climatic zones.

With cycads, the same choices for variety and position are less critical. I have observed most of the varieties growing together under similar circumstances. By that I am referring to this large range of 150 species that I have been able to photograph and observe together. I daresay there would be some of the equatorial African or American varieties, which could require more specialised conditions. If one was able to obtain any of those less procurable species and they were showing signs of distress on planting out, I would suggest their removal to a more sheltered position. Cycads are most hardy to replanting, if carefully handled.

If your climatic zone is within the 35° north and south latitudes and you don't receive heavy frosts or snow, or some other extreme, you should have no problem with any of the cycads in areas receiving some sun or filtered sun to brighter shade. If you find, following a reasonable time after planting out, that the foliage, on the species planted, is sunburning or bleaching badly, I would suggest the plant's removal to a semi-shaded or more shaded position.

Most *Cycas, Dioons* and *Encephalartos* and many of the *Zamias* and *Macrozamias* grow in full sun. This does not mean that they will not grow in more shaded situations enjoyed by the other cycads. Also those other species are happy in all but the most burning of sunny aspects. It may

*Encephalartos manikensis*
This cycad has a thick trunk.

*Microcycas calocoma*
An attractive rare cycad for semi-shaded positions.

*Archontophoenix* sp.
These palms are providing cover for a rainforest setting of ferns and other plants.

*Lepidozamia peroffskyana*
Beautifying a shady corner with ferns and tree ferns.

be best to first check the origins of all the species before making a planting. For plantings in areas receiving full sun the selection of *Cycas, Dioons, Encephalartos* and *Macrozamias* with blue-coloured foliage should prove the hardiest.

Cycads may be used in landscape lay-outs, either in conjunction with palms, or any other plants in a landscaping plan. They may be spot-planted whenever supplies of a different variety can be procured. Because they are much slower in growth than most other plants, they rarely become troublesome. Allow enough room for each to grow larger, without being cramped out by something else. Many of the faster-growing plants alongside of them will probably be gone and replaced by the time the cycad has become a pride of the garden. Also, because cycads, in many cases have vicious leaf spines that can tear flesh or give a nasty puncture, they should be placed back out of the way of traffic areas. This same advice can apply to some of the spiny palms also. If you already have one planted that could be trouble later, move it now. The longer you leave it, the more bloody the task becomes. Give it a "mate" if you can so that you can also enjoy the interesting seed cones and collect fertile seeds. But remember you have to finish up with a male and a female, not two of the same sex.

A garden of cycads can also be quite effective, either on their own or along with cacti. It can be topped off with gravel, woodchips or pine chips to keep down the weeds. This will further eliminate the need of close weeding and injury. A few Mexican or Chinese ornaments are attractive for effect.

## WHY NOT A RAINFOREST

Cycads can be successfully displayed in shaded areas with ferns both large and small. Tree ferns are particularly good for this purpose. In reality, this effectively becomes a home rainforest, which can be used along the shaded side of a house or between two dwellings, where only a few hours of sunlight are received each day. Cycads for mixing in this area are *Lepidozamias, Bowenias, Macrozamias, Ceratozamias, Microcycas* and several of the *Zamias.* Some of these are softer and less dangerous from spike injury. Bark, groundcover logs and fancy stones will add to the effect. When shaded don't forget to add a few small palms like *Chamaedoreas, Licuala, Linospadix,* or even a spiky *Calamus* just to name a few.

In a new garden, you may wish to include a rainforest, as mentioned in an earlier paragraph. Select a position on the warmer side, because so many of the palms originate from the tropical or temperate regions. The area would need to be of sufficient size to achieve the right effect. I would not advise anything below a minimum of 6m (20′) in width, but larger would be better. You could even use the whole yard. A rustic effect can be achieved simply, by using large bush logs as an edging for a car drive, or just to create the bed. These should not be sunk into the ground in heavy soil (clay areas). It is better for the whole bed to be raised by adding soil so as to create a properly drained garden bed, which will get the plants and palms away to a healthy start. I do not believe that plants placed in dug out holes (except in sand, where larger holes should be filled with soil), get away speedily and healthily. If the expense of filling the area with soil is too great I would suggest mound-planting, to keep the plants at their purchased pot-size heights above ground level. These mounds won't be noticed later as the vegetation fills up, or bark or leafmould are added. Ground covers, ferns, *asparagus* and small shrubs will soon hide it anyway. Don't forget a few cycads as well. Top off the planted area with a surface mulch of leaf mould, wood-chips or similar dressing. For economy, purchase a leaf thrasher, to maintain your own mulch material from cut back vegetation.

If the sunny side is exposed to full sun and climate permits, I would suggest a row of shrubs, or a *Philodendron,* a *Schefflera,* or sun-loving palms like Golden Cane, or some other suitable alternative. Behind this you may inter-disperse palms and a few low trees, plus tropical plants, once some height and protection has been achieved by the earlier planting. When the bed is fully shaded you may also add some *Chamaedoreas, Laccaspadix, Linospadix,* etc. until the desired fullness is reached.

Your local climatic conditions must dictate what plants and palms can be used. The above examples are intended for mild climates only. Don't use any of these if they are going to be grown in a cooler zone.

In creating this rainforest, the next essential is the rain, and this can be created with sprinklers. Rather than rely on daily hosing of both the soil and foliage, it would be far better to install a sprinkler system. This should be preferably controlled by a time clock, or one of the inexpensive timing devices available these days. It should be timed to operate for 10 to 30 minutes daily, with maybe 10 to 15 minutes during winter and two 10 to 15 minute intervals on hot summer days. The sprinklers should be capable of spraying some water over the foliage. These days, inexpensive black plastic hose watering systems can be assembled by the most inexperienced person.

*Ptychosperma macarthurii*
In a garden setting with some added colour.

*Veitchia joannis*
In a setting with colourful lower plants.

*Syagrus romanzoffiana*
An advanced specimen in a mound-formed, landscaped bank.

*Encephalartos princeps*
This cycad from rocky habitats has been planted in a rockery.

A rainforest is most rewarding. It is different, and will be greatly admired by visitors, as I know from experience, and is pleasant for warm summer days. You may even wish to install your barbeque there as well, but make sure you are not cooking when the sprinklers are timed to come on.

## PREPARATION AND PLANTING

I am a "crank" on drainage. I believe that digging a hole in hard ground or into clay holes only creates a slow draining water-well that breeds root-rot and fungus and could be the reason behind the slower get-away in growth of a palm or any other plant. I have proven this over and over again, in my own garden, as well as with countless numbers of gardeners over a lifetime of customer problem-shooting. If the palm or cycad you are about to plant has been selected for a wet, or a dry position, or just the average garden position, the planting method is the same. The circumstance is the only difference. If in sand, or built-up gardens, this problem does not apply, as the drainage from the plant is maximum. In fact something would need to be added to hold moisture longer.

If the planting is to be done in a lawn, or flat area, or garden bed that has not already been built up to naturally drain freely, I would suggest the palm be removed from its container, and sat on ground level. With a lawn it would be preferable to remove the turf first and replace this with some soil or potting mix. If required, a concrete ring, or an old car of truck tyre (painted of course), or rock edging, may be placed around the soil of the palm to the height of the plant's root-ball and then to any width that is desired.

The reason for sitting the now un-potted plant on the surface of the ground, is to assess the height of the planting plot. This will become the minimum height of the bed, or planting plot. You may make it to any height you like, simply add soil or potting compost under the pot-ball until you have the desired height. With a bed for multiple plantings, the largest pot-size will govern the ultimate height of the bed. If desired there is one way of reducing any pot-ball height reasonably safely. Once removed from the pot, the lower part of the root-ball may be teased out, by pulling out from under the centre, with both hands, to either side, at the same time. Work your way around the ball, spreading wider, until the desired height is achieved. A certain amount of the soil will fall away from the roots while doing this,

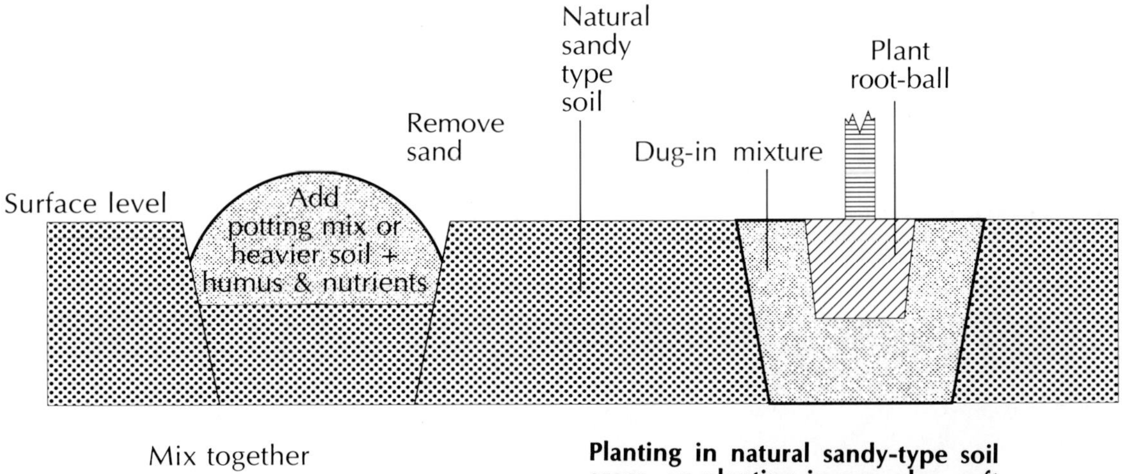


**Planting in natural sandy-type soil areas, or planting in non-clay, soft and pre-dug, raised beds containing garden soil. The lower level of the sketch would indicate the ground level in this case.**

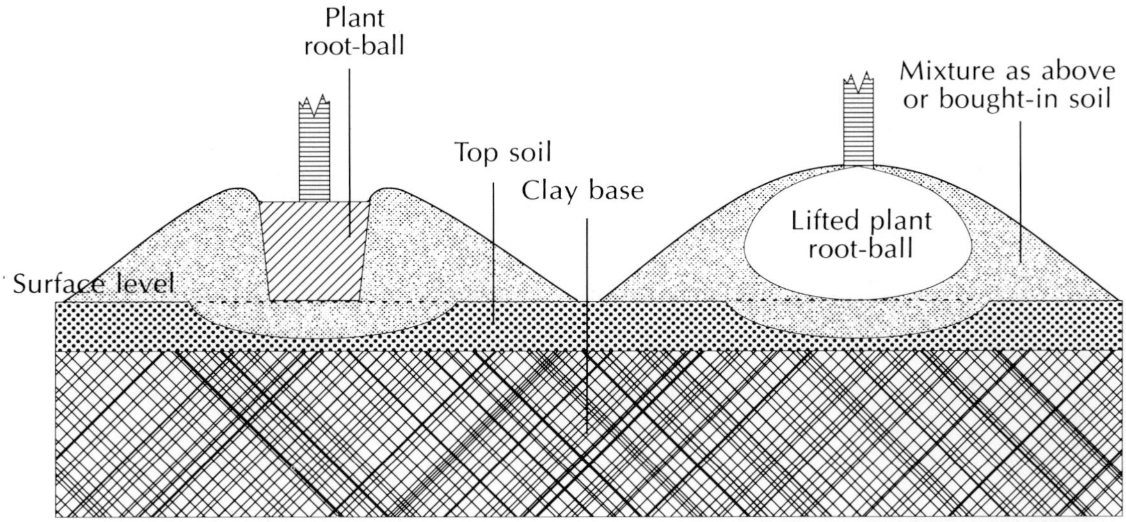


**Non-clay topsoil. If this drains immediately water is applied, it may be dug and mixed with above.**

**Clay. Do not disturb, or dig holes in this. These would not drain quickly or properly and result in root-rots.**

**WRONG PLANTING METHOD for heavier (clayey) soils. This sort of diagram is too often used on plant labels, or in garden books, without mention of soil type, ground form or drainage requirements. Excess water will remain in the hole, causing root problems and diseases.**

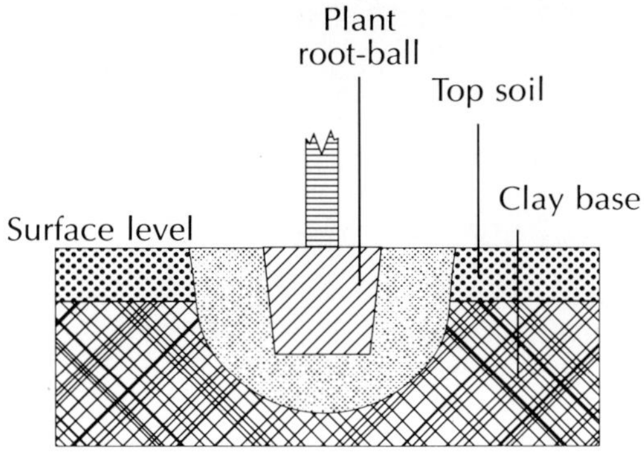

*Archontophoenix alexandrae*
A large lifted plant about to be planted, showing the root-ball.

*Syagrus romanzoffiana*
These large palms were placed on ground level and then a mounded rise was formed around them.

but if planted immediately and watered straight away and regularly, no permanent damage should result. Landscapers plant in this manner all the time. It also helps to spread the roots, if done carefully.

The above is easy in the case of the smaller-sized pots to about 30 cm (12") pots. It becomes a lot harder with advanced or even larger "lifted", or balled plants. However, the advice does not differ on the aspect of placing the plant above drainage levels. As you have probably paid a large price for these plants, it is rather important that you then get the benefit from such added expense. In these cases, it still means a mound of soil around the root-ball, by building higher with stones, or other walling materials, until the desired bed height is achieved. Another method is to form a mound around one or more larger plants, as a bank and which also may be used to incorporate a number of plants or a garden divider. This may even be grassed over, or formed into a rockery, by using rocks on the surface to the desired height. Refer to the picture illustrating this. The other rockery plants can be added in the process. Built-up gardens require regular waterings, or better still, permanent laid-on water equipment, if maximum growth and appearance is to be obtained. Lack of this attention defeats the purpose of obtaining advanced plants, in the first place.

In selecting the soil to fill-in around the palm or cycad, I would suggest if the expense is not too great, that this be a potting mix, either solely or generously mixed with the natural soil, or just closest to the plant. The results will speak for themselves. If not potting mix, then mixtures containing rotten garden compost, animal manures (but not on their own), weathered and composted hardwood sawdust, peatmoss or any other recognised potting soil component. After planting, use an application of complete fertiliser, or one of the time-release fertilisers, may be used. It is then time for a thorough watering. A dribbling hose over a time may penetrate best.

As previously mentioned palms and cycads will grow better and can be relocated better if they receive regular water. This applies to all the species, until they become established in their new position. In the case of the "wet" or rainforest species, regular watering should be kept up during their life. Wetting the foliage also helps. This gives nicer and fuller trunks on palms, as well as lusher foliage. It is interesting to observe some garden palms with lovely plump trunks which have probably been maintained by an attentive home owner. Then when the home either changes ownership, or the owner becomes less diligent, the trunk growth becomes noticeably thinner. A return to consistent watering then only increases the size of the latter grown part and not the earlier part which suffered from the dry spells.

Once the plant is established in a drier area where regular water cannot be maintained, water can be very gradually lessened for the drier location species of palms or cycads. However in that region or during dry times, occasional deep floodings are better than regular surface wetting.

When replanting one of the larger palms, which may have had to be handled by two or more people, or even lifted in by a crane, it is important to observe the diagrams on planting with adequate drainage. Study the picture on planting with a crane. When planting is done, it will be necessary to attach either three or more firm stakes just out from the palm, and fix firm ties from these to the trunk, without damaging it. In the case of very large palms, it may be necessary to use something firmer in the form of heavier ropes and ground pegs to tie these to. Protect the trunk where these attach because they may have to remain for many months, until new roots form, to hold the palm firmly themselves.

*Livistona decipiens*
Ribbon Fan Palm, is quite hardy in most zones.

# PALMS AND CYCADS IN POTS

## PALMS FOR POTS OUTDOORS

Palms are wonderful pot-plants. If given the right-sized containers, quite large specimens can be grown in areas where it is impractical to have garden beds, such as verandas, patios, paved terraces and barbeques, or around swimming pools. In fact any new areas that need dressing up in a hurry, can be quickly decorated. As well as these home uses, there are hundreds of commercial uses for potted palms outdoors. These include beer gardens, car yards, restaurants and shops. In local government they are needed for entrances and even as a decoration near special walks and footpaths, on platforms of railway stations, and around outdoor stages, just to name a few.

Selected specimens should be progressively potted to larger pots, after the roots have fully netted-up the smaller-sized pots. Potting mixes with a peatmoss base would be best for conserving the moisture and prevent it from drying out too rapidly. Surface mulches are also beneficial. But neither of these replaces the need for regular daily waterings. This job can be made easier, where large pots are in fixed positions, by using permanent watering methods which are readily available and quite inexpensive. The kits can be

*Ptychosperma macarthurii*
A lovely cluster palm for warmer climates.

*Licuala grandis*
Ruffled Fan Palm is a dainty palm for tropical climates.

*Dioon spinulosum*
This shapely cycad has a lovely appearance, and will take full sun.

assembled by any inexperienced handyman, using black plastic tubing and push-in fittings. They may be fitted to time clocks, or tap timers. Potted palms should be regularly fed with a complete fertiliser, or one of the time-release fertilisers, to maintain their lush green foliage.

In selecting palms for pot specimens, it will be handy to draw on the information in the quick reference table. It may be of added assistance to remember that the single-trunk palms as they grow taller, offer height, but leave a bareness at the base. To pep-up appearances with these, it may be best to also use a fill-in or spillover, or a few hardy annuals for colour. As the palms are potted larger, so they will also grow taller, with their umbrella of leaves at the top. Taller-growing cluster palms will do the same, but will occasionally grow an extra trunk to retain a filled appearance if well-looked after.

Selections from the medium-sized solitary-trunk palms would react somewhat similarly in growth to that of the taller ones, but not achieve that single trunk as quickly. These could also require a fill-in to cover bareness. Medium-sized cluster palms and fan palms in particular, could achieve a "fuller" appearance for longer. A compact single-trunked palm would be best selected from those which grow their trunk-rings the closest. An example of this is between *Archontophoenix alexandrae* (the rings are closer) and *A. cunninghamiana* (further apart).

In selecting from the smaller range, it must be remembered that many of these are undercover palms and not suitable for hot open-sun situations. They may also be very slow in growing to achieve that filled-up appearance, so it may be best to multiple plant about three plants, nicely spaced in the one pot. There are many species to choose from so it should not be too difficult to achieve a good selection.

Other factors to bear in mind when selecting the climatic zone for pots are position, whether hot/sunny, filtered or partial sun, or no sun. These factors may widen the selection range by including those from a warmer or cooler zone. Also, refer to the other heading categories for added information to help you get a choice which will give you many years of enjoyment.

## CYCADS FOR POTS OUTDOORS

Cycads are very slow-growing, so they make great pot-plants, which you have as beautiful specimens for a very long time. Though slow to reach what might be termed admirable, they become proud treasures, when nicely looked after. The reason they are so slow, is that they usually only grow their new leaves once a year. Sometimes, if damaged (in nature this may be by a bushfire), they may grow a second batch of leaves so you must assure these are not damaged to remain as beautiful specimens. New soft growth hardens up to resist further problems, and is rarely damaged after that, unless by physical means. The smaller juvenile cycads may even have a few re-growths a year, but matured specimens are usually restricted to one.

Most of the varieties of cycads finish up with their matured leaves dark green, but some have new growth coloured from silver to bluish, or even yellowish. This adds greater variety in making a selection. Another thing to remember about cycads is that they are usually armed with sharp leaf points. Most have either a sharp point or edge spines to the leaflets, or both, while a few have only a varying number of thorns on the leaf stems or none at all. Don't let this put you off them though, just place them out of a traffic area. The most popular cycad on earth is *Cycas revoluta,* and this has sharp tips.

Cycads are treated in much the same way as potted palms so that previous information is also applicable here. Cycads are sun hardy as well as total shade subjects, and those areas in between. They are extremely hardy even in pots that may dry out for short periods, but I am not recommending this course. You only get out of a thing, what you put into it. Care for them and you will be rewarded, by better appearance.

As previously written, there are not as many recorded facts as one would wish, even though they are one of the world's earliest plants. But any range offered for selection would more than likely be mentioned in this book. Most are very hardy and their growing conditions will not vary greatly. With potted plants it is a simple matter to move the pot, should any problem arise. I feel any abnormality showing up would more than likely come from a bad potting mixture, bad drainage, lack of water, too much sun, lack of nutrients or an imbalance of these, or even scale insects (one of their only insect problems). All of these can be easily rectified.

Remember a male and female of a species can make them still more interesting pot-plants by producing fertile seeds.

## PALMS FOR INDOOR DECORATION

This subject is more fully dealt with throughout the book, each variety having a "Position Indoors" heading, followed by individual advice for that variety. However, there are a number of conditions that are applicable to all palms being used as indoor plants.

*Zamia fischerii* (compact form)
This species makes a compact potted cycad. Plant a male and a female together.

*Zamia loddigesii*
*Zamias* are compact, lasting pot-plants.

*Chrysalidocarpus lutescens*
A popular private and commercial indoor palm.

*Howea forsteriana*
Top commercial indoor palm, worldwide

**POSITION:** Always select a position with the brightest natural light possible. For most palms this would be better if it had warmth, as well. These add to the health and life of the plant, if it is to remain there for very long. Such factors may be of the utmost benefit to those touch-and-go varieties. The hardier varieties may tolerate slightly darker positions for a while, but also will respond better with the brighter conditions. Always avoid drafts and strong breezes. Remember air conditioning dries out the air, and can be harmful to humid-loving varieties. Regularly hose down their foliage, if possible, to maintain the moisture levels and also to cleanse them.

**WATERING:** Palms need to be kept damp and not allowed to dry out. The types of palms that are most adaptable for indoor use are generally from rainforests. Rarely are dry-country palms used for this purpose. Excess water should not remain in saucers though as this would only sour the soil and lead to root-rot. Saucers would be best filled with pebbles or similar, to stand the pot on. This way, it becomes impossible to have the pot standing in water. The wet pebbles supply a constant source of humidity, to further add to the health of the palm. An occasional hosing-off of the foliage is beneficial, as previously mentioned.

**FEEDING:** Occasionally use some complete fertiliser, or one of the time-release fertilisers (Osmocote or Nutricote). The latter comes with time releasing over periods of 3 months or 6 months. Excesses, or too frequent fertilising can be harmful. The best time to use a fertiliser is when the plant is most likely to use it. That is when the palm is making growth, usually the warmer months. Excess salts from continual use of fertilisers, can be flushed out, by taking the plant outdoors on dull or wet rainy days, for a good hosing. Don't forget it though or it may get sunburnt, and the foliage will be spoiled, by sun and wind.

**OTHER HINTS:** Lay off the use of White Oil, for leaf shining, it can burn or kill. Look for insects. If you see some, a stiff hosing will often move them. Scale on palms can be rubbed off with a soapy cloth to save spraying or by using a mild White Oil spray if the first method is unsuccessful. Re-pot into larger pots only when you need to. They don't require large pots under indoor conditions. It may be better to report in the same pot, using some fresh indoor compost. Refer to re-potting section.

It is a good practice to rotate your indoor palms, allowing those in darker positions to recuperate, in the more favourable positions.

Whenever you take your palms outdoors, or into the shower for a hose-off, follow this by wetting down the foliage with a water solvent foliage fertiliser. One such brand is Aquasol, but every country will have its own range of brand names.

## CYCADS FOR INDOOR DECORATION

This is a relatively new addition to the indoor range of plants and is only now being marketed. At present, little information can be supplied about a range of successful varieties. As a result, I will have to leave this to you because the only way to find out is to try them.

There are some facts that will help you when you do take a cycad indoors:

* Always make sure that the new seasons growth has hardened up, before taking it inside;
* Likewise take the plant back outdoors again for it to add its new season leaves. This is usually spring, so have it back onto a veranda-type setting, perhaps early spring, guarding it from sunburn, until it has acclimatised again;
* In many cases the plant given a bit more time under better light outdoors, may produce more leaves than if it were kept in shade, or indoors;
* Upon being taken indoors the cycad foliage will pale slightly and the centre crown of the plant will be more open and attractive. The stems will lengthen a little and depending on the degree of natural light, the plant could lose one or more leaves over the season period. The colour will darken again when taken back outdoors. Rotate around by turning, to eliminate "leaning" and misshaping;
* Cycads respond to gentle and irregular fertilising indoors. A good practice would be an occasional hosing or showering to remove leaf dust, followed by a mild application of water soluble foliage fertilisers.
* Do not use white oil as a foliage shine; the leaves are naturally glossy enough. White Oil should only be applied if being attacked by scale insects. A hosing will maintain the sheen, by removing the dust;
* Recent tests carried out indicate that cycads are very worthwhile indoor specimens, but so far only two varieties of *Cycas* and *Zamia* have been experimented with (see reference below);
* Leaf-making on young cycads is more erratic and could commence at any time. Should this begin indoors, put them outdoors immediately to avoid damage, also then protect them from sun damage.

*Chamaedorea seifrizii*
Reed Palm is wonderful indoors, given good natural light.

*Zamia furfuracea*
Jamaican Sago Tree has stood up to indoor testing.

*Ceratozamia 'Palma Solo'*
An attractive new release which looks good in a pot.

*Cycas circinalis*
Fern Palm has shown good results indoors, but grows best with good natural light.

Other than the above data, cycads should be cared for in accordance with the information given in the previous palm section. They are well worth a trial and could be most rewarding.

**(*) Reference to an article appearing in *Palms and Cycads of Australia,* magazine no. 7, April 1985, by Dehgan & McConnell.**

# POTTING PALMS AND CYCADS

## POTTING MIXES — BOUGHT OR HOME-MADE?

There are many good packaged potting composts available in garden outlets and nurseries. These rarely contain soil, owing to the scarcity of this valuable commodity around our cities these days. Instead the mixes utilise waste materials which are differently composed in the various localities and enriched by the addition of proper nutrients. Most of today's progressive nurseries are using these commercial mixes because of their consistently even growth results. Many, otherwise wasted compounds, such as aged or composted wood sawdust, grounded pinebark, peanut shells, sugar cane waste, scorea, along with much of the humus-materials, are mixed with sand, peatmoss, vermiculite, perlite, rice hulls and nutrients, to form a balanced mix of even consistency in each batch every time it is formulated.

It is important, when purchasing a potting mix, not to go for just the cheapest, with the thought that some standards-authority is looking after your interests, because they aren't. Some can be cheap and nasty. Make sure the one you purchase contains some peatmoss, or comes highly recommended, by a previous user.

Some of the indoor potting mixes are a little too flaky for palms and cycads. It is advisable to add a percentage of extra soil to these, but don't make the mix over-heavy and spoil the drainage quality. Both palms and cycads prefer an open or free-draining growing medium, which will retain some moisture, but it must not be soggy. Only change those potting mixes that need changing. It is also a good idea to refer back to the text of each species, to ascertain any special needs.

Should you wish to make up a mixture yourself, you may find you have many of the ingredients at hand, either the ones mentioned or good substitutes. The following recipe should make a reasonable mix. The measure will be based on the standard 9 litre (2 gallon) household bucket. For larger quantities, do your sums.

- 1 bucket of good garden soil, as used in the vegetable garden, or 2 to 3 of sandy loam to replace the soil and sand;
- 1 bucket of salt-free sand to be added, unless using the sandy loam;
- 1 to 2 buckets of mixed fine and coarse humus in the form of peatmoss, rotted compost, old sawdust and pine-bark, or similar products blended together.

At this stage mix all the ingredients and ascertain the texture for good drainage. This must have an openness, while retaining a good moisture-holding quality. It is sometimes a good idea to have some commercial potting mixture to compare with, if you are unsure or mixing-up for the first time, so as not to continually repeat mistakes. If this is alright continue; otherwise add the necessary ingredients to make the mixture acceptable.

- Add 200g (7–8oz) Dolomite, or Garden Lime 250g (8–9oz) if a bucket of peatmoss was used, as this is more acid.
- Add 250g (8–9oz) complete fertiliser, containing mineral trace elements. If these are not included add the trace elements separately. If possible calculate the pH and make suitable for a pH of around 5.5 to 6.5, before mixing the amount of lime. This pH is only average, not specific. If the palms or cycads originate from limestone areas more lime may be needed to raise the pH to between 6.5 and 7.00.

Thoroughly mix through immediately, but if the mixture is for storage and later use, only add the fertilisers when about to use them. If the mixture is dry and crumbly, moisten and mix this through before you commence potting.

## FIRST POTTING

This would apply to your seedlings or older plants being lifted and potted for the first time. In this instance the potting mix should not contain very much fertiliser or fresh manure.

*Chamaedorea seifrizii*
Carefully remove the soil and cut rooted divisions off for re-potting. New sucker sections would be preferred.

*Cycas armstrongii*
A long-lasting pot-plant.

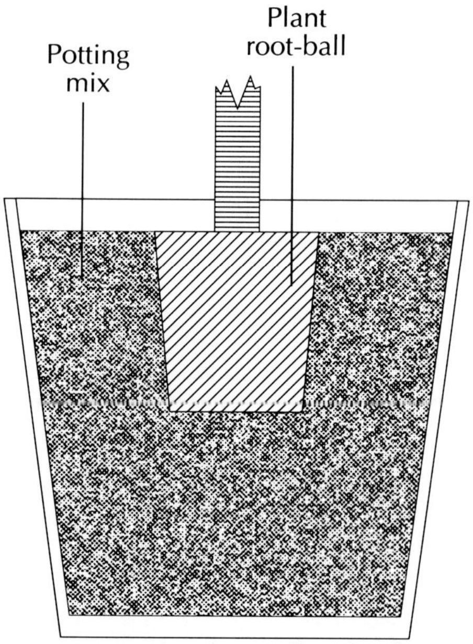

Potting-up from a smaller sized pot into a larger one, but only one or two sizes larger.

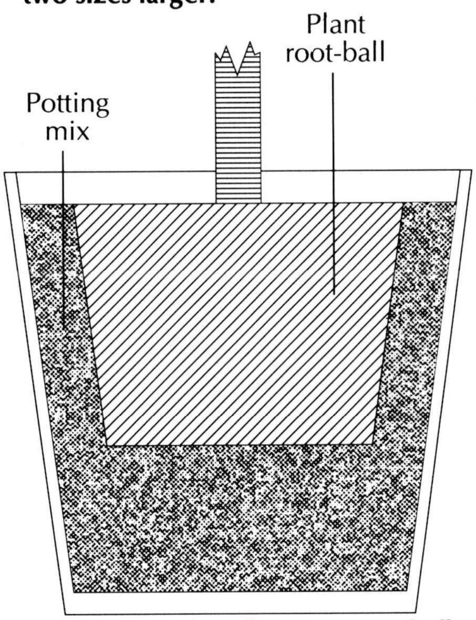

Re-potting a shrunken pot root-ball into the same sized pot.

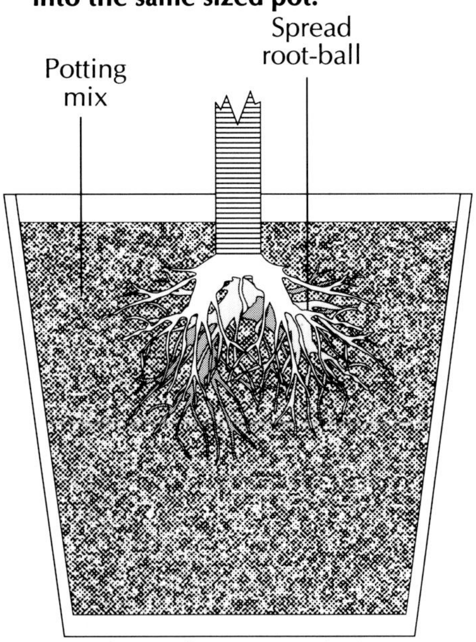

Spreading the roots before re-potting into a larger sized pot.

Cut these to a quarter to a half of those mentioned. Feeding can follow later, when the plant is showing signs of needing the food, by growth-making. Also a weak solution of liquid, foliage-absorbing fertilisers can be applied. Too much soil nutrient could impair the formation of new roots by burning these off.

Select small pots for first potting, the smaller that can be practically used the better. The idea is to allow the roots to net-up in these pots quickly, so they can then be potted-on in a better mixture to make growth. Make sure to keep up the water daily.

This may be deemed as the infant stage, with intensive care needed, away from the hot sun and winds. A bush house may be best.

## POTTING ON OR RE-POTTING

This will apply to all sizes whether re-potting in the same-sized pot, or potting up into a larger-sized pot. The potting mixture to be used will be with full fertiliser added.

Re-potting may be necessary because the soil in the pot has shrunken down, well below the rim of the pot. Don't just top it up. Remove the plant from the pot by inverting and tapping it out. Judge the depth required to bring the root-ball to the desired height and fill this amount of new potting compost into the bottom of the pot. Place the plant on this and then fill around it with the new potting compost, firming this gently down. An application of time-release fertiliser may be applied to the root-ball. Water gently until penetrated.

When potting on to a larger-size pot, it is important to remember to only advance one or two sizes larger at any one time. It is a mistake, to think small plants will grow quicker by going into larger pots immediately. The growth size doesn't seem to begin until the roots have netted to the outer edge of the pot. Also, re-potting shouldn't be done before that size is fully netted, roots begin to appear through the pot holes.

As with re-potting, place potting mix in the new pot, first assessing the depth the root-ball will be placed on this, at the desired height. Then fill in the sides and tamp this down. Some thoughts do exist on opening out the root-ball, from the centre-base with both hands, to the sides and widening this out. This may be done if desired. It is particularly useful if potting into less-tall, wider pots. Top off with some time-release fertiliser unless you have disturbed the ball too much.

Should this have happened, then play it cautiously and leave the fertiliser until after a few weeks have elapsed. Water thoroughly.

## PARTIAL RE-POTTING

This is generally practised with larger pots, after a considerable time has elapsed for netting up and/or if it may be thought advisable to change the potting mix for a fresher one. In most cases it would be too heavy or too inconvenient to move these around, so the job is tackled where they stand.

This is carried out systematically over a certain time period in two, three, or four stages, and can take up to a year or a number of years. *Re-potting in two stages.* Mark the surface soil in the pot into two opposing segments, each about one-third in from the outer edge, but well short of the trunk. Then use a trowel to clear out these areas by cutting down through the roots and soil.

Replace with fresh potting mix, tamping down and thoroughly watering the old root-ball as well as the fresh compost. Make sure to keep the old root-ball adequately wet while the roots are re-growing into the fresh area.

The opposite sides of the untouched area may be removed the following year but only if those renewed sections have fully recovered and netted-up again with new root growth. *Re-potting in three stages.* This is done a third at a time, over a 3-year period. It could be harmful to tackle the next removal too soon; wait until the previous area has grown before continuing.

If tackling this method with cycads, it would be advisable to hold up the foliage first, wrapping this in an old sheet, to avoid injury. This also applies when moving a larger specimen, to protect the eyes, as well as the arms.

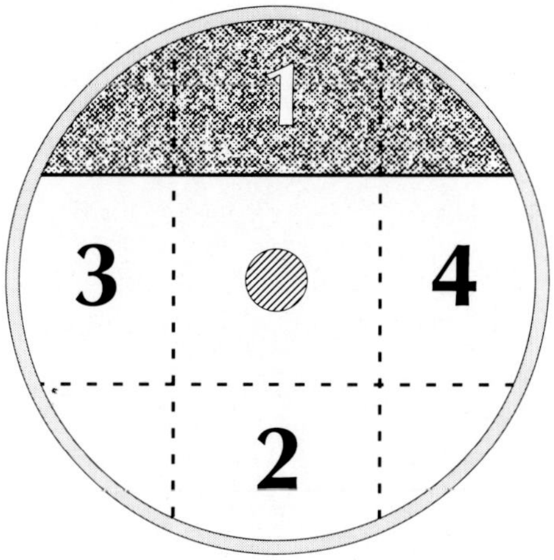


**FOR LARGER TOO-HEAVY POTS**
Remove one or two opposite soil and root sections. Next season repeat in the untouched sections. Alternatively remove 1, 2, 3 & 4 over four seasons.

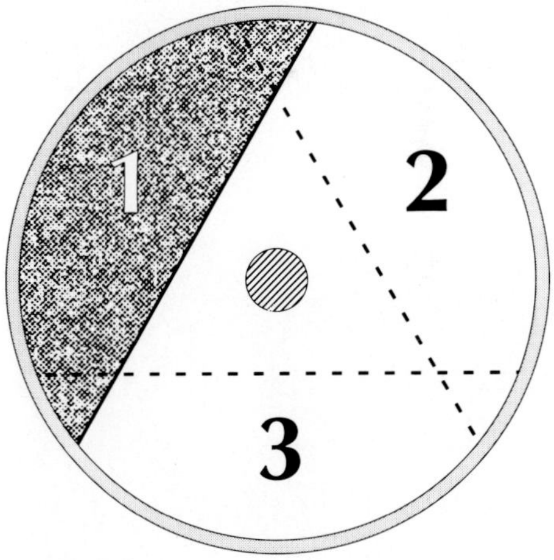


If preferred remove 1, 2 & 3 over three seasons allowing for regrowth of roots between.

# PROPAGATION

Both palms and cycads are reproduced from fertile seeds. Some cluster palms may also be multiplied by the division of a clump, making sure that the section to be removed has its own set of roots. Cycads, or some varieties of cycads, may also be multiplied by division, as well as by rooting of large cut sections of their trunks.

*Rhapis humilis*
Slender Lady Palm may be carefully divided to increase numbers.

*Encephalartos trispinosus*
Seeds are being germinated on the potting soil surface.

# PALM PROPAGATION

This is dealt with individually with each variety in their text. However, there are a few points that should be taken into account before commencing. When produced, the seeds have a fruit covering. In nature, this is an attraction to animals, to eat off, or swallow. In that way they then carry that seed further afield before it can germinate and commence to grow. Masses of seeds do fall off under the palm and sometimes do germinate there, but it is strange how they rarely grow up to compete with the parent palm.

For quick removal of the outer fruit, nurseries generally place the coloured mature fruits in a container with some water. This soon "pickles", and after a week or so, the rotted fruits come away from the seeds. These are washed and planted right away, or dried for despatch.

Before planting seed, first find its zone of origin and its natural habitat (rainforest, or sandy scrub, or dry plain, or whatever). These facts are useful to obtain top results in the planting of the seed, plus the covering and the best position. For instance a nursery would most likely raise palms from a tropical region under heat and keep them quite moist. The laypeople wouldn't have the equipment for this, but in the nursery a plastic bag could be used. All he has to do is place some pre-moistened peaty potting mix (preferably sterilised first) to a ¼ or ⅓ level in a large plastic bag, add the seeds and only just cover or half cover these with the damp mix. It shouldn't require any attention as the moisture is sealed inside, but it shouldn't be allowed to dry out and would be too hot exposed to sunlight. Hanging in a shady tree is a good place. Alternatively the bag could be stood, staking first to hold it upright, against a shaded or early sun wall. The degree of early sun governs the heat. Not too much or you'll cook the seeds. Should you only have a few seeds of a number of varieties, you may prefer to plant each species in a pot and place all these in a plastic hanging or standing bag. Don't forget to label each.

With open-growing palms from a temperate region, good results have been obtained from the above method also. Another method which has shown good easy results, is to place potting mix, or a good soil mix, into a polystyrene fruit box to about half way, plant the seeds with a light covering (equal to the seed-size) and then top off the remainder of the box with dried grass mowings, loose leaf mould, straw, or similar. Make sure to keep up the watering. The box may be in semi-shade sun or shade, according to the locality of the seeds being raised.

After germination takes place, the young palms should be transplanted at a young age, before root damage from handling could occur. They must not be planted into too-large pots. That may slow down the roots from netting up quickly, or they will stand still instead of growing. It is better to move these up through the sizes, beginning with pots from 50–100mm (2″–4″) if they will fit so as not to crush the roots, or appropriate-sized or shaped pots that will serve minimally. Water early after potting and maintain this liberally. The young palms would be best grown under the protection of a bush house, or the appropriate protection needed for that variety, at the time of the year to keep them on the move. This protection could vary according to climate.

Clumping palms, such as *Raphis* or *Chamaedoreas*, may also be carefully divided, to obtain another plant. This is best practised with potted palms, where the plant may be removed from the pot and the soil washed off the roots to enable the situation to be carefully assessed. Make sure there are good roots on each section and cut cleanly, without damage to the remainder. Pot the sections into appropriately smaller-sized pots to net up, using a low fertilizer potting mix. They may be potted on at a later date, into a richer potting mix. Weekly solutions of mild foliar fertilisers would help to maintain good appearance, while re-establishing the plants.

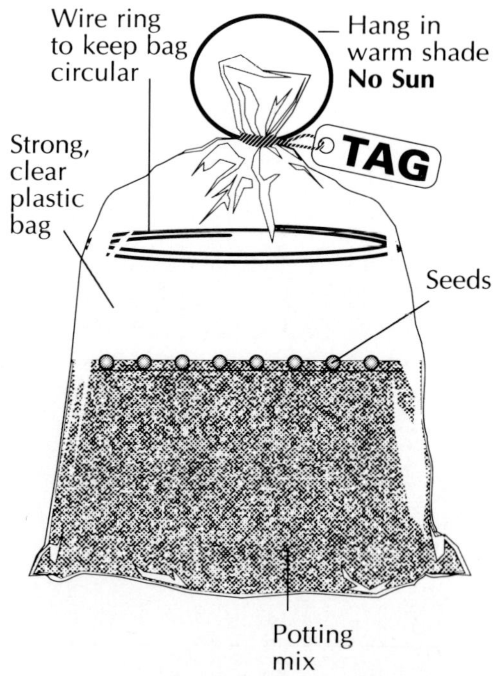


**Seeds may be planted in a clear plastic bag and placed or hung away from the sun's direct rays.**

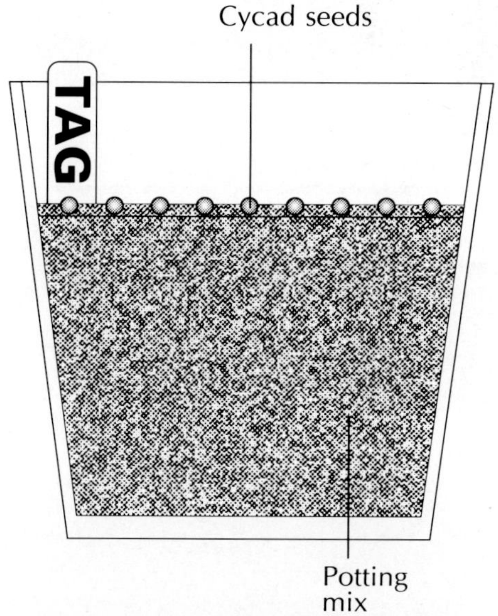


**Seeds may be planted in a polystyrene fruit box or similar drained box.**

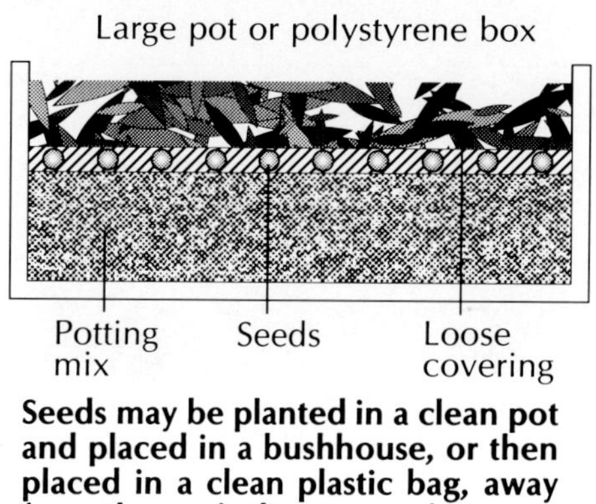


**Seeds may be planted in a clean pot and placed in a bushhouse, or then placed in a clean plastic bag, away from the sun's direct rays, for more tropical varieties needing higher humidity.**

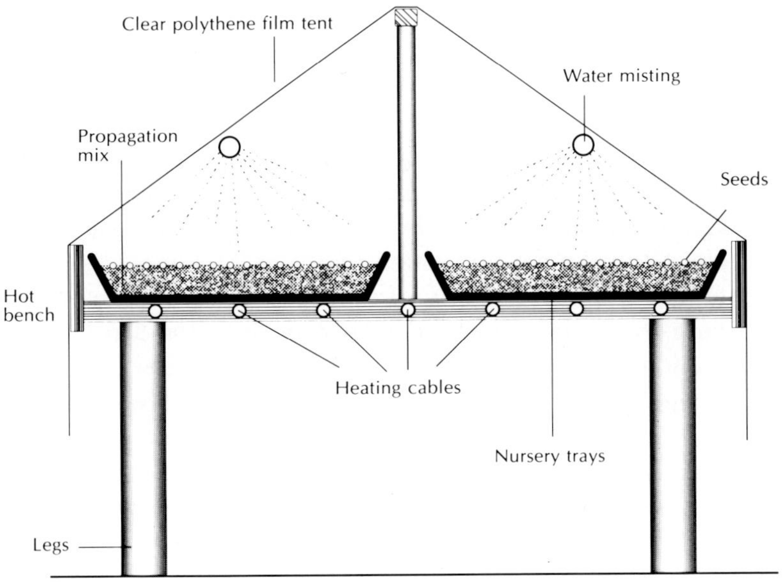


**This hot-bench raises seeds quickly.**

*Encephalartos manikensis*
Divisions may be carefully separated if each is rooted.

*Zamia portoricensis*
These types of clumps may be carefully separated.

# CYCAD PROPAGATION

All varieties of cycads are grown from fertile seeds, or the nuts produced in the F cones. Differing from palms, which have flowers to become pollinated and then produce a fruit and fertile seed, cycads produce a nut or infertile seed, regardless of pollination. That is why all seeds obtained need to be tested for viability. This can be ascertained by a water test, the fertile ones sink, or by sacrificing a seed or two if they're plentiful, by cutting and opening them to see whether the embryo shoot has formed. This is indicated by a coiled suspensor at the tip. On older 6 to 7 month seeds the radicle root may even be pushing its way out.

Plant the seeds, about half covered, into a bed or pot of potting mix, keep watered and wait until the shoots appear. This may be many months. Pot singly at this stage, so as not to cause later damage to the roots. Do not remove the nut, as the young plant will still be drawing nourishment from this. Allow it to drop off in its own time. When potted, and for many years while in the juvenile stage, young cycads should be potted on, but kept under shade protection, or a bush house, or glasshouse for the more tender species. They are many years getting to maturity.

Another method of obtaining some species of cycads, is by off-shoots. These come on older plants of *Cycas, Encephalartos* and occasionally other cycads. These suckers attach around the base and before removing these, carefully remove the soil from that vicinity so as not to damage any root system. Only remove the off-shoot if there are roots attached. Dust the "cuts" later with sulphur or another fungicidal dust. Leave exposed for a week to allow drying out, thus avoiding rot.

A further method of vegetative propagation is by the growing of a large limb. It is a good idea to water well, a few days before cutting this trunk, to allow the storage of water. Cut off cleanly and dust the cuts with fungicidal dust, against rot. Allow the cut-off section to dry, at least one week. Prop up the trunk over a bed of sand, it may rot if planted, but here it should callus and take root, into the sand. Water the foliage and trunk daily to avoid dehydration.

The advantage of this vegetative form of propagation is that the plant would be of the same sex as the parent, whereas seeds are of unknown sex and it takes a long time to ascertain the sex of seedlings. Divisions can be an advantage between friends who have older plants of opposite sex. The re-establishment process is still very slow, though, as is everything to do with cycads.

For further information on propagation, refer to Len Butt's article 'Cycads in Australia', as well as the text for each variety.

## PESTS AND DISEASES

Palms and cycads do suffer from a few problems in the way of insect attack and, very occasionally, some diseases. Other than apparent leaf-chewing, the other problems are rarely seen. Most of these problems are confined to palms, with their softer and more edible leaves. Some pest-control advice will be offered, but as each country has different sets of regulations on chemicals available, you may be better advised to contact your own nursery for local recommendations.

*Caterpillars* have eaten these leaves. *Grasshopper* damage is similar.

## PEST CONTROL

*Caterpillars* are chewing insects and these may be located at the back of the palm leaf, where they will attach leaflets together and harbour out of sight between eating as well as forming into their moth cycle to multiply and attack again if unchecked. *Control* Searching for these is generally too late to avoid damage and only possible with small plants. Spray all the foliage with Maldison (Malathion), or Carbaryl, or some other effective insecticide.

*Grasshoppers* are also troublesome here, but are very difficult to kill with the weaker insecticides allowable. Catch these larger specimens during the cooler hours, when they are still sluggish. The newly hatched youngsters will be controlled by the sprays mentioned above.

*Scale insects* on both palms and cycads, will be *controlled* by cover spraying with a white oil solution, during a cool period. Also more effective are the combination of White Oil and Malathion.

*Mites and red spider mites* are sucking insects, showing noticeable leaf damage and yellowing. These are particularly evident with indoor specimens, away from natural predators (ladybirds). *Control* Miticides, Folimat, Kelthane, sulphurs, etc. Be sure to use a follow-up spray in 10 days to catch the re-hatching. If re-infestation still appears, change to a different Miticide in case there is an immunity to those previously used. I also like to take the plant outdoors and give it a thorough jet-hosing to wash those pests away, and I avoid spraying with chemicals where I can, but at times this can't be avoided.

*Thrips* generally hatch during hot spring weather and attack in plague masses, on palms and other plants. *Control* Dieldrin may be banned as are most of the longer-lasting insecticides. Use of Metasystox, Maldison, Folimat, or similar, will control these, in that absence.

*Mealy Bug* is hard to kill, and hides out of sight. The white fluff down the trunk reveals its presence.

*Scale insects* may be pink, white, brown or black. This type is the hardest to kill.

*Scale insects* with *black smut* residue are thickly infesting this *Zamia angustifolia*

*Mealy Bug* is a curse to control. It is accompanied by white fluffy deposits around its environs. With palms, it will harbour within the rough leaf bases and even around the roots or any hidden areas. It is a sucking insect. *Control* Use any method to flush them out or kill. Popular methods are hose water-jets, probes or brushes in order to physically remove as many as possible, before spraying. Find the best insecticide available and spray this into all the hidden areas. Folimat is one of the only chemicals to kill these. Sometimes though, like similar products, it may only be effective for short periods, so it is necessary to repeat when re-infestation is observed.

## DISEASES AND CONTROL

*Fungal diseases* There are several fungal diseases and these are usually indicated by brown marks and dead patches in the palm leaves. They can even cause the death of the palm in severe cases. *Control* Treat by cover spraying with a fungicide, such as Dacanyl, Mancozeb or a similar recommendation from your nursery.

*Brown spots* may also appear on some varieties as a result of attempting to grow a warmer climate species in too cold a climate. This would appear during winter and early spring, but disappear during the warmer months. It may pay to replant this palm during late spring to a warmer protected position, if not too large.

*Lethal yellowing* has no control. With this distress which becomes most obvious, the palm loses the fruit, then the lower leaves yellow and fall. This condition works upward until all the leaves yellow and then the palm dies. If it is a problem in your area, remove and burn the palm. Do not plant another in the same position.

*Cycad yellowing* is a condition which occasionally occurs with some varieties. It is not altogether sure what causes it and it seems to not continue in the new growth. Remove the affected leaves, perhaps by doing so the plants may regrow some fresh ones. Yellowing may also be caused by incorrect pH of the soil to suit that variety. Do a pH test, and check up on the variety, in case there is a pH preference mentioned.

The above are the most apparent problems that will appear in both palms and cycads. Should any problems other than these occur, submit a sample of the affected foliage to your local nursery or agricultural department officer for advice.

# PALM AND CYCAD SOCIETIES

Should your interest in palms or cycads reach an advanced stage I would suggest you contact one of the many world-wide associations that now exist.

There are a growing number of these now operating, in many countries, including Australia, Great Britain, USA and South Africa. The easiest way of assessing a contact would be to approach one of the botanical gardens' botanists, who, I feel sure, would point you in the right direction.

These societies have regular publications to assist and keep you up-to-date with the latest on both of these popular plants. They also often have field excursions and market days when you can inexpensively add to your collection, with otherwise unobtainable stock, as well as gain lots of ideas. Membership could be inexpensive and most helpful.

Some of these societies are:

*Australia*
Palm & Cycad Society of Australia,
PO Box 1134, MILTON, Queensland 4064.

*Great Britain* has a Threatened Plants Committee
Threatened Plants Committee,
Botanical Gardens Conservation Co-ordinating Body,
C/- The Herbarium, Royal Botanical Gardens,
KEW, Richmond, Surrey, England.

*South Africa*
The Cycad Society of South Africa,
20 Maryvale Road, WESTFIELD, 3630

*USA & World Association*
The Palm Society Inc.,
Box 368, LAWRENCE, Kansas 66044, USA.

*Red spider* or *Two-spotted Mites* are minute, but the damage they cause is evident as fine cream spotting.

*Cold marks* on a palm leaf indicates that the species requires a warmer climate.

# EVERLASTING PALM

This is the lastest from the Americas. It needs little attention, no watering, no spells outdoors, just an occasional dusting, to keep it looking good. It stays as you see it, a permanent indoor palm.

It is in fact a reconstructed *Washingtonia* palm. The parts are collected from the desert, dried out, permanently coloured and then reconstructed into this lifelike model.

These days other forms are being assembled to represent other palm species.

# PALMS

The varieties are arranged alphabetically.

*Archontophoenix alexandrae* This large group is magnificent at Flecker Botanical Gardens, Cairns.

*Acoelorrhaphe wrightii*
Everglades Palm.

*Acrocomia aculeata*
Macaw Palm or Gru Gru.

# ACANTHOPHOENIX

One variable species, *A.rubra,* from Mauritius and Réunion Islands, now rare.
A solitary trunk feather leaf palm.

# ACOELORRAPHE

One clustered species, native to the brackish swamps of Florida, the Caribbean Islands and surrounding parts of Central America.
*A. wrightii,* Everglades Palm, Silver Saw Palmetto. This palm has slender multiple trunks, forming a dense clump 3–10m (10′–30′) high. The green fan leaves are silver beneath, and usually accompanied by many dead leaves retained on the plant. The creamy inflorescences extend from the top of the plant and mature to small round black fruits 10mm(3⁄8″) in diameter.
**Culture and habitat** A tough but moisture-loving palm, growing slowly if inadequately watered. A large grower, it is best for parks or large gardens. It is suited both to the tropics and to coastal temperate areas, with fine specimens growing in both Cairns (tropical) and Sydney (temperate) botanic gardens.
**Sex** Hermaphrodite. Self pollinating with one plant.
**Position indoors** Not known to be used indoors.
**Propagation** Fresh seeds usually germinate within 2 months, occasionally slower. The juvenile leaves are undivided.

# ACROCOMIA

About 26 solitary species, located in the West Indies and from Mexico south to Argentina, in drained savanna and grazing lands.
*A. aculeata,* Macaw Palm (Dominica, Martinique). A handsome palm, at first glance resembling *Syagrus romanzoffiana (Cocos plumosa),* but armed with black spines on the 10–12m (30′–40′) trunk, as well as on the feathery all-green leaves. The spent leaves fall, leaving a clean but thorny trunk. Juvenile plants are also noticeably thorny. The pendulous inflorescences later bear round, green to yellow-brown fruits 4cm (1.5″) long.
*A. media* (Puerto Rico). A handsome palm with a 15m (50′) trunk covered with black spines. Similar to *A. aculeata,* but the leaves are silvery beneath. Fruits are 4cm (1.5″) long and yellow.
*A. mokayayba* (Brazil). This species differs from the preceding two in that the trunk has few spines and the old leaf bases remain attached to the trunk for a long period.
**Culture and habitat** These palms are as yet little grown in commercial nurseries, but widely cultivated in botanic gardens. Acrocomias are growing well in Sydney and Brisbane botanic gardens (to 33°S) as well as botanic gardens in their native tropics. They require a frost-free, warm situation and regular watering, as well as good drainage.
**Sex** Monoecious. Self pollinating with one plant.

**Position indoors** No reports, but rather thorny for an indoor plant.
**Propagation** It may be best to file the tough seed coat or soak the seeds for 2–3 weeks, keeping the water warm. They should germinate in 4–6 months. Young plants may need cold protection and shelter until hardened off or planting out.

## ACTINOKENTIA

Two solitary species from New Caledonia, growing in the understorey of wet forests on slopes from 60 to 1000m (200′–3000′)
*A. divaricata* (syn. *Kentiopsis divaricata*). A small palm with a slender 3m (10′) trunk topped by a yellowish crownshaft and only about four feathery leaves on red stalks. The newly opened leaf is red. Small branched flower spikes produce oblong fruits 25mm (1″) long.
*A. huerlimannii* is the other species.
**Culture and habitat** This small palm has been a popular collector's palm around the world for over a hundred years, often being grown in glasshouses as a potted palm. Commercially it is rare, and it is only likely to be seen in the larger botanic gardens. When planted in a garden, it requires a shaded or undercover situation, with a mineral-rich, slightly acid soil. A surface mulch would be an advantage, as well as regular watering. It appears that tropical, subtropical and warm temperate climates are suitable for outdoor plantings, but protection is needed from cold winds. In other zones the protection of a glasshouse or similar structure would be needed.
**Sex** Monoecious. Self pollinating with one plant.
**Position indoors** Appears to be a good subject, but no precise information is available.
**Propagation** No germination times are available.

## ACTINORHYTIS

Two solitary species, one native to New Guinea and the Solomon Islands, growing in tropical rainforests from sea-level to 1000m (3000′). The other is restricted to the Solomon Islands.
*A. calapparia* (New Guinea). This palm has a slender trunk to 12m (40′) high topped with a crownshaft and a small array of feathery glossy green fronds. The 6cm (2″–2.5″) long fruits are oval and red at maturity.
**Culture and habitat** Rarely seen out of botanic gardens elsewhere, this palm is becoming popular in South-East Asian countries, as a magic medicine plant. The seeds are also chewed as a betel nut substitute. They should be able to be grown successfully outdoors in the tropics, but would need more protection in cooler areas than this. They require regular watering.
**Sex** Monoecious. Self pollinating with one plant.
**Position indoors** No information available on indoor use.
**Propagation** No information available on germination times.

*Acrocomia media*
An attractive tall palm.

*Acrocomia mokayayba*
In this species the trunk is unarmed.

*Aiphanes corallina*
This attractive palm stands out in a group.

*Aiphanes erosa* (juvenile)
Macaw Palm.

## AIPHANES

About 38 solitary species, occurring from the West Indies to northern South America, and most abundant in Colombia.

*A. acanthophylla* (Puerto Rico). This palm has a slender black-spiky trunk to 8–10m (25′–35′). The feathery leaves have irregular leaflets, and are thorny above and below. Fruits are 2cm (¾″) in diameter, round and bright red.

*A. caryotifolia,* Coyure Palm (Colombia, Venezuela, Ecuador). This species has a ringed trunk to 10m (30′) high covered in long black spines, and an open crown of broad, wavy leaves on pale, spiky stalks. The spiny inflorescence of yellow flowers is followed by round bright red fruits 15–20mm (½″–¾″) long.

*A. corallina* (Martinique). A handsome species bearing spines on the trunk and leaf stalks. It is now endangered in the wild.

*A. erosa,* Macaw Palm (Barbados). This species has a densely spined trunk to 4–5m (12′–16′) high. This is topped with a compact head of nicely arranged bright green feathery leaves on stalks which are both spiny and white-scurfy. The crowded head of perfumed cream flowers is followed by round red fruits 2cm (¾″) long.

*A. lindeniana* (Colombia). A slender palm, much resembling *A. acanthophylla,* from the mountain regions of Colombia.

**Culture and habitat** *Aiphanes* species are quite striking in appearance and suitable only for warm to tropical climates. They grow in jungle rainforests which have high rainfall and humidity. Planted as garden specimens they would require the shade and protection of other plants, in a selected warm area or suntrap. Because of their vicious spines, they should be placed in areas where they will not be accidentally bumped into. When planting, dig in some compost or potting mix, and hose the plant down daily to maintain humidity. For pot use, pot on as required, using an indoor potting mix. Best pot sizes are from 200–500mm (8″–20″). *Aiphanes* require shade and warmth, with constant watering. In cooler climates a glasshouse or warm greenhouse would be best in the early stages.

**Sex** Monoecious. Self pollinating, with one plant.

**Position indoors** These are interesting palms for indoors, with their black spines. Select a sun-warm corner, or a sunroom. Take out and hose down regularly but do not leave in the sun. Water regularly and feed occasionally to keep green.

**Propagation** Seeds germinate in 1–2 months. Grow in a warm, sheltered, shady position, or in a glasshouse.

## ALLAGOPTERA

Five usually clustered species, native to Brazil and Paraguay, as sea-coast sand dwellers and in scrub further inland.

*A. arenaria* (syn. *Diplothemium maritimum*) (Brazil).

*Actinokentia divaricata*
A popular small palm.

*Aiphanes acanthophylla*
Note the rings of black spines on the trunk. Ruffle Palm.

*Actinorhytis calapparia*
A magic medicine plant in Asia.

*Aiphanes caryotifolia*
Coyure Palm. Attractive but spiky. The fruits are eaten.

*Aiphanes lindeniana*
The trunk has black spines.

*Allagoptera arenaria*
This palm is suited to sandy seashores.

A small clustering palm 1–2m (3′–7′) high. Its bushy growth covers the ground with waxy, dull fronds. The yellow flowers on erect stems are followed by oval green, edible fruits 2–2.5cm (¾″–1″) long. Salt-loving, suitable for seaside planting.
**Culture and habitat** These ground-hugging dwarf cluster palms originate from the Bahia area of Brazil. The sandy soil is poor and they resist the salt winds. They should be particularly useful in gardens near to ocean fronts where a hardy dwarf palm is required either as a garden specimen or as a potted palm. I have seen them used in coastal or near-coastal locations to 35° latitude, around Sydney and Brisbane.
**Sex** Monoecious. Self pollinating with one plant.
**Position indoors** No information on indoor use.
**Propagation** Seeds germinate within 3 months.

## ALLOSCHMIDIA

One species, *A. glabrata,* from New Caledonia.
A solitary trunk feather palm.

## ALSMITHIA

One species, *A. longipes,* from Taveuni in the Fiji Islands.
A solitary trunk feather leaf palm.

## AMMANDRA

One species known, *A. decasperma,* a second discovered, from Colombia.
A solitary trunk fan leaf palm.

## ANTONGILIA

One species *A. perieri,* separated from *Neodypsis,* but still doubtful.
Solitary trunk, feather leaf palm.

## ARCHONTOPHOENIX

Two solitary species from eastern Australia, from forested areas, swamps and rainforests in lowlands, on mountainous slopes and in valleys, growing either singly or in groups.
*A. alexandrae,* Alexandra Palm (central to northern coastal Queensland, Australia). The trunk to 25m (80′) high is thicker and has closer rings than that of *A. cunninghamiana.* The crownshaft is bright green, and the leaves are also bright green above, and silver below. The handsome flower skirt has cream flowers, followed by round fruits 2mm (½″) wide, red at maturity. *A. alexandrae* variations are not authenticated at this stage, but are distinctly different from the pure species, and include:

*A. alexandrae var. beatricae* resembles *A. Kuranda* except that the swollen, bulbous trunk has distinctive 'steps' at the rings.

*A. alexandrae 'Kuranda'*, from the Kuranda area of North Queensland. This seed is widely used by many of the nurseries, producing an attractive broad-trunked potted palm. The older plant is also very appealing, producing a swollen trunk topped with a broad crown of leaves. It later produces a broad skirt array of fruiting stems below the crownshaft. There are many fine specimens around Queensland's Gold Coast.

*A.* sp. 'Mount Lewis'. New species allied to *A. alexandrae*, is distinctly different in both juvenile and mature forms. The juvenile leaves are stiffer, broader and paler green in colour and distinctly silver beneath. As they mature the crownshaft turns rusty brown. The leaf segments are broader and thicker than in *A. alexandrae*. The inflorescence is similar to that of *A. alexandrae* and originates from the Mount Lewis area of North Queensland.

*A. cunninghamiana*, Bangalow Palm, Piccabeen Palm (coastal regions from central New South Wales to Queensland, Australia). The trunk grows to 25m (80') high. The crownshaft is yellow-green to brownish. The erect leaf fronds have paler stalks and leaves which are green above and beneath, but also have sparse brown scales beneath. The spikes of flowers are mauve, and the green fruits 10–15mm (½") long, later turn red.

*A. cunninghamiana* (variation used in some nurseries) Features a thicker trunk, closer rings and more compact growth, with brownish leaf stalks. It is much like *A. alexandrae* in stature, but bears all the other characteristics of *A. cunninghamiana*. Grown by some nurseries in Australia from their own seed stocks.

**Culture and habitat** These handsome and graceful palms are most people's concept of what a palm should look like. Their popularity with the buying public finds them a ready market through chain stores and other fast-moving trade outlets. In Australia they are the number one seller on the palm market. If you have purchased an *Archontophoenix* it is very important for you to understand a little more about this beautiful palm, if you wish your Alexandra or Bangalow Palm to survive.

In nature, these palms are mostly to be found in the rainforest areas of the north-eastern coast of Australia. *A. cunninghamiana* is more cold-tolerant than *A. alexandrae*, and stands of this palm may be seen in the rainforests today as far south as Batemans Bay in New South Wales, whereas *A. alexandrae* would be seen at its best in the rainforest around Cairns in North Queensland. However, in sheltered gardens in temperate Sydney, both varieties may be seen growing to perfection.

The important message to be learned from their natural rainforest habitat is that *Archontophoenix* species require regular wetting. If they get this, along with an abundance of water to the roots, they will be quite happy in your garden, even in open competition with shrubs and trees planted quite close to them. They will also look more handsome with

*Archontophoenix alexandrae*
Alexander Palm, King Palm.

*Archontophoenix alexandrae 'Kuranda'*
From the Kuranda area of Queensland.

*Archontophoenix* sp. *'Mount Lewis'*
Note the colouration of the crownshaft.

*Archontophoenix alexandrae* var. *beatricae*
This palm has a stepped bulged trunk base.

a regular once or twice weekly wetting down of the foliage, using a foliage fertiliser such as Aquasol. Ensure that you use the recommended strength though, and spray when the day has cooled off.

*Archontophoenix* palms may be seen at their best around Queensland's Gold Coast, where many strains of this attractive palm are being marketed. Graceful palms have become the garden symbol of the area, and provide a relaxed tropical appearance that suggests an idyllic existence. Many of the hybrid strains are being marketed there, and these have more pleasing characteristics than the original species found in nature.

**In the garden** In areas sheltered from strong winds, *Archontophoenix* will tolerate sun, or grow equally well in shade or under the cover of trees. They require plenty of water more frequently than most other plants. This is important for good appearance. They even thrive in constantly wet areas. When the leaf grubs are active, a few foliage sprayings with Carbaryl or other more effective insecticides will generally clean these up. *Archontophoenixes* may be planted as single specimens or in stands of three about a metre apart. However, these days they are being marketed with clusters of three to five plants in a pot. Do not separate these, but plant them as they are, and you will end up with a very handsome stand of attractively trunked palms.

**Sex** Monoecious. Self pollinating with one plant.

**Position indoors** To be happy indoors, *archontophoenixes* require very high humidity, more than the average home can provide. Without this, they tend to brown off toward the tips of the leaves. I suggest that they be taken outdoors a couple of times per week and hosed down, to keep the leaves in good condition. Place in the brightest position possible, even near an open window, with filtered, but not burning, sunlight.

**Propagation** Propagation is from seeds, which are easily grown, but are quicker with heat, usually taking 6–12 weeks to germinate. Fallen seed often germinates beneath the palm, once the fruit has rotted away, if watering is kept up. Pot up as required, only one size larger at a time.

## ARECA

Sixty solitary or clustered species, distributed from India to southern China and southward through Malesia to New Guinea and the Solomons, growing as undercover palms in tropical rainforests.

*A. aliceae* (Australia?). Originated from a plant located near Cairns in North Queensland, but not known in the wild. It has clustered green bamboo-like trunks to 3m (10′) or more high and bearing bright green feather leaves. The inflorescences whisk broom-like and bearing 20–25mm (1″) fruits.

The authenticity of this species is disputed by some who think it may at sometime have been introduced. It is nevertheless being cultivated in many areas of the world as *A. aliceae.*

*A. catechu,* Betel Nut Palm (origin uncertain, but

*Archontophoenix cunninghamiana*
Piccabeen or Bangalow Palm.

*Areca aliceae*
Reportedly of Australian origin, but much resembles *A. triandra.*

*Archontophoenix cunninghamiana*
This sturdier form is used commercially by some nurseries.

*Areca catechu*
Betel Nut Palm, at Batu Caves, Malaysia.

*Areca concinna* (juvenile)
A clumping palm for shady spots.

*Areca guppyana*
A small grower with a single trunk.
(Photo C. Migliaccio)

likely Malaysia). A slender palm with a single trunk to 10m (30′), with deep green feather leaves. The fragrant white flowers are followed by orange to scarlet fruits 50mm (2″) long containing seeds known as betel nuts. An industry has evolved in Asia, supplying betel nuts to the numerous people who chew them for their intoxicating properties.

*A. concinna* (Sri Lanka). A multiple-trunked species, very similar to *A. triandra,* but with scarlet fruits containing seeds which are also used as betel nuts in Sri Lanka.

*A. guppyana* (Solomon Islands). A slender single-trunked species only growing to about 3m (10′) tall and with stilt roots at the base. There is a crownshaft, topped by a small head of leaves, each frond only having about five pairs of large segments. The seeds within the bright red fruits are sometimes used as betel nuts.

*A. ipot* (Philippines). Resembling *A. catechu,* but smaller in growth, only reaching about 4m (13′) tall, with a ringed trunk topped by a swollen green crownshaft. The green feathery leaves are gracefully displayed, as is the heavy cluster of red fruits 5cm (2″) long.

grown in the Rare Plants House at Fairchild Gardens in Florida. No other information is available, except for its description in the picture and that it does produce viable seeds occasionally.

*A. triandra* (India, Malaysia). A clustered palm with many slender trunks to 3m (10′) high. The fragrant light flowers are followed by orange fruits 20–25mm (1″) long. This species prefers shade.

*A. vestiaria* (syn. *A. langloisiana*) (Sulawesi). A colourful multiple-trunked species. The ringed trunks from 5–8m (15′–25′) tall are topped with reddish-orange crownshafts and glossy green feathery fronds. The short yellow to orange inflorescence is succeeded by orange fruits 2cm (¾″) long.

**Culture and habitat** *Areca* species are single- or multiple-trunked understorey palms from tropical rainforests and the various species differ in their habitat requirements.

*Areca* products are widely used by the local people for their medical and narcotic values, as well as for decoration.

**In the garden** Suitable only for the tropics or subtropics, or suntrap areas elsewhere. Select a warm position, sheltered from cold winds, and a well-drained soil. Keep well watered.

**In pots** These make attractive potted palms and will tolerate sun or shade, provided the position is sheltered and warm. Water frequently and do not allow to dry out.

**Sex** Monoecious. Self pollinating with one plant.

**Position indoors** *Areca* species, like those of *Chrysalidocarpus,* have been widely used indoors for many years. They are very dainty, and well worth a try.

**Propagation** Seeds, which must be fresh, germinate in 1–3 months. Keep in a warm place, as they are very cold-tender. Bottom heat is beneficial.

# ARENGA

17 solitary or clustered species, distributed from eastern India to southern China and southwards through to northern Australia. They grow in tropical rainforest in the lowlands and hills, the smaller species being understorey plants.

*A. ambong* (Philippines). This palm has a short branching trunk with a clump of attractive spreading fronds resembling *Caryota*. The leaflets are green above and silvery beneath.

*A. australasica* (coastal North Queensland). This tall, clumping feather palm has up to three main trunks surrounded by smaller suckers. Each trunk dies after seeding, and may then be removed to give a tidier appearance.

*A. caudata* (syn. *Didymosperma caudatum* (Malaysia, Indonesia). A small clumping palm with multiple slender trunks to 2m (6′) forming a thick clump. The leaves have thin stalks and the leaflets are well spaced and irregularly wedge-shaped, resembling *Caryota*. The flowers are borne amongst the leaf bases, and the fruit is oval, 15mm (½″) long, and purplish.

*A. engleri* (Taiwan). A dwarf, clumping feather palm with short trunks. The dainty notched leaves are whitish underneath. This species will grow well in cooler temperate climates and will even tolerate light frosts. The fruits are carried in large clusters and turn red at maturity.

*A. fastuosa.* A lesser-known species being tried out

*Areca ipot*
A small grower for shady areas.

*Areca cf. multifida*
A rare species, at Fairchild Gardens, Florida, USA.

*Areca triandra*
A tall clumping species.

*Areca vestiaria*
A colourful clumping species.

*Arenga ambong*
The caryotoid leaves are silver beneath.

*Arenga australasica*
This species forms a medium clump.

*Arenga caudata*
This species was previously known as *Didymosperma caudata.*

at the Royal Botanic Gardens, Sydney. No further information is available.
*A. listeri* (Christmas Island). This single-trunked species grows to medium height. It prefers a warm climate and shade. It is currently under trial in the Royal Botanic Gardens, Sydney. After years of seeding, the plant dies.
*A. microcarpa* (New Guinea). A clumping species to 7m (20′–25′) tall. The broad feathery fronds are deep green above and silvery beneath, and form a dense cover. Flowers are deep purple. Suited to the tropics.
*A. mindorensis* (Indonesia) This juvenile plant is being grown under trial at Sydney Botanic Gardens.
*A. obtusifolia* (Java). A clumping species of medium height. The dark green leaflets are silver beneath. May be dioecious.
*A. pinnata* (syn. *A. saccharifera*), Sugar Palm (East Indies) A large feather palm on a trunk of 10–15m (30′–50′) high and about 30cm (12″) in diameter, with the old leaf bases remaining attached and covered with long black spines. The arching leaves are borne on stout stalks from a crown, and the numerous leaflets are deep green above and whitish beneath. The inflorescences are 2–3m (6′–10′) long, and from these the sugary sap may be extracted over a period of time. The 3.5cm (1½″) fruit is yellow and brown. After many years of seeding, the tree eventually dies.
*A. tremula* (Philippines). A trunkless clumping feather palm to 2–3m (6′–10′) with narrower green leaves. This species will tolerate temperate climates, doing quite well in areas similar to Sydney.
*A. undulatifolia* (Borneo). Very eye-catching as a specimen, or when the silver undersides of the unusual wavy, broad blue-green leaflets are seen from beneath the tree. The single or multiple trunks are fibrous and the leaf stalks are dark. Suited to the tropics.
**Culture and habitat** These tropical palms have occupied an important place in tropical village life through the centuries. Their sap has yielded palm sugar and an alcoholic drink, the trunk and leaves are used in hut building, and the fibre from the trunks makes ropes.

The species of *Arenga* vary from tall single-trunked forms to trees with multiple trunks, and even trunkless species that more resemble shrubs in appearance. *Arenga* is closely allied to *Caryota* and *Wallichia*, with a number of similar features. There is an irritant in the seeds, and each trunk will die after bearing its final seed crop, at the age of perhaps 12–15 years. In the case of the single-trunked species this is the end of the plant, but multiple-trunked species lose only one trunk at a time.
**In the garden** Arengas are quite hardy in tropical gardens as well as in many warm temperate areas, where they require a deep rich soil, ample moisture and protection from cold winds. They are not as hardy in cooler temperate areas, so if you want to grow them there, a protected suntrap position should be found and some compost or potting mix added to make the soil richer and deeper before planting. Hose the plant daily to increase humidity.
**In pots** *Arenga* species are sun hardy, but prefer semishade for best appearance. Repot upwards as

*Arenga engleri*
A handsome medium-clumping palm.

*Arenga microcarpa*
Note the handsome feather leaves.

*Arenga tremula*
**A clumping feather palm.**

*Astrocaryum alatum*
**A species of Star Nut Palm.**

*Arenga undulatifolia*
**A very eye-catching species.**

*Astrocaryum mexicanum*
**Star Nut Palm.**

required. Best pot sizes are from 200–500mm (8″–20″) (also smaller sizes indoors). In cooler climates *Arengas* are best grown in glasshouses or conservatories.

**Sex** Monoecious. Self pollinating with one plant.

**Position indoors** Select a warm position near a bright window, or a sunroom. Keep moist and hose down the foliage frequently, for added humidity. Remove outdoors again at first signs of distress.

**Propagation** Seeds of most varieties germinate better with bottom heat and will germinate within 2 months, but some take longer (up to 12 months). Growth is rapid. Clustered species may also be propagated by division if not too large. Make sure divisions have roots.

## ASTEROGYNE

Three species, from Central America and northern South America.

Usually solitary trunk and with leaves unsplit.

## ASTROCARYUM

47 solitary species, distributed from Mexico to Bolivia and Brazil, also in Trinidad. They grow as understorey plants in lowland rainforests, except for the more light-demanding species, which grow in well-lit open forests.

*Arenga obtusifolia*
This species forms a large clump.

*Arenga pinnata*
Sugar Palm. A massive grower.

*Arenga mindorensis* (juvenile)
Under trial at Sydney Botanical Gardens.

*Astrocaryum mexicanum*
An unusual inflorescence. Note the fruits in the pouch (right).

*Attalea crassispatha*
This species is closely allied to *Orbignya*.

*A. alatum* (northern South America). A slender species, less spiny and with smaller inflorescences than *A. mexicanum*. The leaflets are silver beneath.
*A. mexicanum* (Mexico and Central America). Another interesting small collector's palm suitable for warmer climates. It has a slim trunk to 2m (6′–7′) covered in nasty spines. The leaf stalks are also spiny, the feathery leaves being deep green above and silver beneath. The interest is in the brown fruiting pouch, and the spiny 4cm (1½″) long oval fruits.
**Culture and habitat** *Astrocaryum* species are natives of Trinidad and surrounding Central and South American areas. They are unusual feather palms, armed with spines, and with solitary or clumping trunks. Their fruits are prickly with interesting markings, hence some are called 'star-nut palms'. Being tropical, they would require a warm position and ample, regular watering. Outside their natural range, these palms are mainly confined to botanic gardens, but an ardent collector will occasionally grow one as a novelty. In Australia *A. mexicanum* is mostly grown. It appears to do well in tropical and subtropical areas, growing out in full sun.
**Sex** Monoecious. Self pollinating with one plant.
**Position indoors** There is no information available on indoor use.
**Propagation** Seeds are difficult to germinate. They should be cracked or soaked in warm water before planting and will then take 1–2 months to germinate.

## ATTALEA

22 solitary species distributed from Panama to Brazil and Peru, in both tropical rainforests and drier forests.
*A. crassispatha* (Haiti). Now thought not to belong in the genus *Attalea*, but not yet placed elsewhere, this species is closely allied to *Orbignya* and will sometimes be found listed in that genus also. It has a sturdy trunk topped with large feather fronds with massed narrow deep green leaflets.
*A.* sp. (at Fairchild Gardens). A handsome sturdy grower with long feather leaves held upright. The narrow deep green leaflets are tightly massed. Note the sausage-like inflorescences. Fruits are oval and about 9cm (3½″) long.
**Culture and habitat** Attaleas are closely allied with *Orbignyas*, differing only in minor botanical features. They are a source of oil and medicines in their native countries. Originating from humid tropical regions, they would require similar conditions for satisfactory culture elsewhere. Refer to *Orbignya* for further details.
**Sex** Monoecious. Self pollinating with one plant.
**Position indoors** Not known to be used indoors.
**Propagation** The seeds must be fresh and should germinate in 2 months.

## BACTRIS

239 solitary or clustered species, located from Mexico and the West Indies to Brazil and Paraguay, with the

greatest density in Brazil. They occupy a wide range of habitats, from lowland and upland rainforest to the fringes of mangrove swamps.

*B. gasipaes,* Peach Palm (Central and South America). This species has a clump of multiple stems, which are noticeably ringed and spiky and topped with a crown of deep green feather leaves. In tropical countries this palm is grown for its attractive yellow, peach-like fruits which hang in clusters; these can be eaten after cooking.

*B. major,* Lata Palm (Central and South America). This palm has thin multiple brown trunks, to 3–4m (10′–13′) and armed with black spines at the rings. These later disappear to leave the trunk smooth and white-ringed. The short inflorescence is followed by purple fruits 5cm (2″) long, which are edible, and are also made into an alcoholic drink. There is an open head of leaves with spiny stalks.

*B. mexicana* (Mexico). The numerous slender spiny trunks of this species bear deep shiny green feather leaves whose stalks are grey with brown spines.

*B. plumeriana* (West Indies). In this species the clustered trunks are thicker than in some others, and are green with brown spines. Fronds are relatively long.

**Culture and habitat** Although *Bactris* is a large genus, the species are rarely grown out of their natural habitats, other than in botanic gardens. They require warm climates and originate from rainforests, either near saltwater estuaries, as in the case of *B. major,* or further up into the mountains.

*Bactris* are rather spiky palms and should be placed

*Bactris mexicana*
The orange-yellow fruits are attractive.

*Attalea* sp. (at Fairchild Gardens, Florida, USA.)
Note the unusual inflorescences.

*Bactris major*
Lata Palm.

*Bactris plumeriana*
This species has heavier trunks.

*Balaka seemannii*
Spear Palm.
(Photo C. Migliaccio)

out of the way of pedestrian traffic, to avoid injury from the vicious trunk and leaf spikes. While tropical, they may grow satisfactorily in warmer temperate climates in coastal regions or in suntrap positions which receive no frost or excessive cold.
**Sex** Monoecious. Self pollinating with one plant.
**Position indoors** Not known to be used indoors.
**Propagation** Seeds must be fresh and are best germinated with bottom heat, taking about 2 months.

## BALAKA

Seven solitary species, native to Fiji and Samoa, in rainforests to an elevation of 1000m (3000′).
*B. leprosa* (Fiji). A small, unusual palm, having a slender trunk 2–3cm (1″) wide which is reddish brown and scaly and up to 2m (6′) in height. The *Caryota*-like, fishtail green leaflets are shiny and have a dented surface. The inflorescence is below the leaves and later bears attractive, oblong, shiny red fruits about 15cm (6″) long.
*B. seemannii,* Spear Palm (Fiji). This palm has a slender trunk to 2 m (6′) tall with a crownshaft and has attractive, unusually shaped leaf fronds.
*B.* sp. aff. *siliensis,* Maniuniu. The trunk is slightly thicker than in the above two species and the feathered fronds have cut off ends.
**Culture and habitat** *Balaka* species come from deep in the rainforests of Fiji, and Samoa. The species differ in the different island, with those from Samoa having more squarely cut leaf ends. The 2m (6′) long slender trunks were prized by other islanders for making spears. At present little is known about growing *Balaka* species away from the tropics, or about their value as indoor or garden palms. They would probably require a tropical climate, or in temperate areas a warm and sheltered shady position, or a heated glasshouse. They come from damp, mossy forests; in areas of heavy rainfall, so any attempt at growing *Balaka* would need to match these conditions.
**Sex** Monoecious. Self pollinating with one plant.
**Position indoors** These are untried as indoor palms, but being allied to *Ptychospermas,* which are excellent indoors, they may be expected to perform well. They are well worth a try if stocks are available, but make sure you give them a little more attention than more common genera, including a daily hosing to keep up humidity.
**Propagation** Propagated from seeds. Little information is available on the raising of *Balaka* seeds, but bottom heat and a humus-rich mix would probably be needed.

## BARCELLA

One species *B. odora,* from Brazil previously *Elaeis.* Solitary trunk, feather leaf palm.

# BASSELINIA

11 solitary or clustered species, native to New Caledonia, in shale-soil rainforests.
*B. eriostachys* (Noumea). This palm grows in a clump, with slender stems to 4m (12′) high topped with a colourful crownshaft shading through red, blue and purple. This is topped with a head of slender feathery leaves. The flowers and the small red or black fruits 10–12mm (½″) long are borne on inflorescences emerging from below the crownshaft.
**Culture and habitat** *Basselinia* species are inhabitants of New Caledonia, where they grow under the shade of other rainforest trees. They are little grown outside their natural habitats except in botanic gardens; therefore information on their culture here is scant. They would be a good, small collector's palm.
**Sex** Monoecious. Self pollinating with one plant.
**Position indoors** Not known to be used indoors.
**Propagation** No germination times available for the seeds.

# BECCARIOPHOENIX

One species, *B. madagascariensis,* from Madagascar. Solitary trunk, feather leaf palm.

# BENTINCKIA

Two solitary species, one native to Nicobar Island and the other southern India, growing in hilly regions, up to 1000–1400m (3000′–4500′).

*B. condapanna* (South India). This species is similar to *B. nicobarica* (see below) but much more slender.
*B. nicobarica,* Bentinckia Palm (Nicobar Islands). This warm-climate feather palm is very similar in appearance to *Archontophoenix.* It has a slender trunk 15–20m (50′–65′) high, topped by a sparse head of deep green feather leaves 2m (6′) long. The fruits resemble red to black cherries, and are 12–14mm (½″) long. *B. nicobarica* grows at lower altitudes than *B. condapanna.*
**Culture and habitat** These palms are native to India and Nicobar Island and are little known outside botanic gardens. There are some specimens growing in both the Cairns (tropical) and the Brisbane (subtropical) botanic gardens, as well as others in many other botanic gardens of the world.
**In the garden** Select a warm position with an ample water supply. Protect from cooling winds, and water regularly. *Bentinckias,* once established, are rapid growers.
**Sex** Monoecious. Self pollinating with one plant.
**Position indoors** No information is available about indoor use.
**Propagation** Seeds germinate within 2 months. Early leaves are broad with two segments for the first year, then part into leaflets to become feathered.

*Balaka* sp. aff. *siliensis*
Maniuniu, growing in Hawaii.

*Basselinia eriostachys*
Note the hanging tassel inflorescence.

*Bentinckia nicobarica*
Bentinckia Palm.

*Bismarckia nobilis*
Bismarck Palm. A heavy fan palm.

## BISMARCKIA

One solitary species, from Madagascar, growing in drier savanna country on the western side.

*B. nobilis,* Bismarck Palm (Madagascar). This tall, heavy fan palm is a striking and impressive plant. The large, very heavy blue-green leaves, well displayed and up to 3m (10′) across, are usually broader than long.

They divide into segments from the leaf base and are carried on blue-green stalks streaked with a white bloom. The trunk is stout and attains a height of about 20m (60′). This palm is fast-growing and self-cleaning. It carries its flowers amongst the leaves and the brown, plum-like fruits are 4cm (1½″) long.

**Culture and habitat** A palm from drier tropical regions and suited to the tropics and subtropics, this palm should also be successful in the warm temperate zone if grown in mild frost-free areas. Because of its great height, it can hardly be protected at its mature size by other plants. It should also be attractive as a potted palm if grown in suntrap conditions for a number of years.

**Sex** Dioecious, requiring both a male and female plant for pollination.

**Position indoors** There is no record of its use indoors.

**Propagation** It has been suggested in some circles that the seeds should be removed from their shells for better germination, and planted into deep pots, so as not to damage the 30 cm (12″) taproot. They should then germinate within 1 to 2 months.

## BORASSODENDRON

Two species, from southern Thailand and northern Malaysia.

Solitary trunk, fan-leaf palm.

## BORASSUS

Seven very similar solitary species, distributed in Africa and east Arabia and Madagascar, around the southern fringe of Asia, and also in South-East Asia including Indonesia. Their habitats range from low sandy plains at sea level to wind exposed mountainsides at elevations to 800m (2500′)

*B. aethiopum* (tropical Africa). A handsome stiff fan palm with a sturdy trunk. The stalks of the large midgreen fan leaves are thick and heavy. It resembles *Sabal.*

*B. flabellifer,* Palmyra Palm (India to Malaysia). This large, heavy-trunked palm grows to 25m (80′) with large stiff fan leaves 2–3m (6′–10′) across on thick prickly stalks; the leaves remaining attached for some time after they die. Flowers are borne among the leaves, and the fruits, which resemble coconuts, are 12–20cm (5″–8″) long.

**Culture and habitat** *Borassus* species are as important in their part of the world as the Coconut Palm: they provide timber and thatching as well as food in the form of the fruits and sugar obtained by tapping the fruit stems. The sugary sap is also used to make alcoholic drinks.

*Borassus* palms prefer the sandy soils near the tropical coast, but appear to be hardy enough to also grow in heavier soils to 30° north and south, as long as their taproot is not damaged. It is best to start off a plant where it is to be grown, rather than take the risk of damage and loss from transplanting.

**Sex** Dioecious. Requires a male and female plant for seed production.

**Position indoors** There is no information on use indoors.

**Propagation** Seeds may require heat or hot water treatment to germinate, and take about 6 months. The shoot first grows downwards for about 1m (3′), then returns to the surface over a period of about 5 months. (The local people regard this shoot as a delicacy.) If these palms are grown in containers, the pots must be deep enough not to damage the emerging root.

## BRAHEA

16 mostly solitary species, located in Baja California, Guadalupe Island, Mexico, Guatemala and El Salvador on limestone slopes and on outcrops in dry areas.

*B. brandegeei* San Jose Hesper Palm (Baja California). A taller-trunked species, to 3m (10′), with the spent leaf bases attached to all but the lower trunk. The 1–1.5m (3′–5′) leaves on 1m (3′) stalks are whitish beneath. Tiny flowers are borne amongst the leaves and 1cm (¼″–½″) fruits are brown with paler mottling.

*B. armata,* Blue Hesper Palm (Baja California). This palm has a thick trunk to 10–12m (30′–36′) covered with dead leaves. The leaf stalks are armed with white teeth, and the heavily segmented fan leaves are blue-grey and waxy. The 5m (15′) inflorescences are greyish-white and protrude beyond the leaves. The brown fruit is marked.

*B. decumbens* (Mexico). A very eye-catching species. The blue fan leaves with bluish-green stalks are displayed upright and with a prostrate growth habit.

*B. dulcis,* Rock Palm (Mexico). A handsome blue-green fan palm with nicely displayed leaves. The fruits are edible and the plant is quite hardy.

*B. edulis,* Guadalupe Palm (Guadalupe Island). The stout 9m (30′) trunk is thick at the base and smooth. The 1m (3′) fan leaves are green both sides and have a mostly smooth stalk 1.5m (5′) long. Flowering stems are long and bear sweet black edible fruits 25mm (1″) long.

*B. edulis* x *brandegeei.* I have seen this interesting hybrid growing in numerous botanic gardens. *B. edulis* has green leaves and grows under 10m (30′) while *B. brandegeei* grows as tall as 25m (75′) and has leaves white underneath.

*B. nitida* (Mexico, Guatemala). The trunk of this species is sturdy and the mid-green, almost circular fan leaves are pleasingly displayed.

*B. salvadorensis* (Central America). This species also has a sturdy trunk. The bright green, almost circular

***Borassus aethiopum***
**A sturdy, heavy palm.**

***Borassus flabellifer***
**Palmyra Palm.**

*Brahea brandegeei*
San. Jose Hesper Palm.

*Brahea edulis*
Guadalupe Palm. The fruits are edible.

*Brahea armata*
Blue Hesper Palm. Note the long inflorescences.

*Brahea edulis* x *brandegeei*
An attractive cross.

leaves are well displayed and have thick armed stalks.
**Culture and habitat** Braheas are large, slow-growing fan palms with a thick, solitary trunk that grow in dry limestone areas, often on hillsides and gullies. They do equally well in tropical and temperate areas.
**In the garden** These palms are suitable for exposed positions in spacious gardens or parklands, and specimens can be seen in most botanic gardens. Select a well-drained position, and water only when necessary.
**Sex** Hermaphrodite. Self pollinating for seed production.
**Position indoors** There is no information about use indoors.
**Propagation** Seed germinates easily, but takes 2 to 4 months.

## BRASSIOPHOENIX

Two species, from New Guinea.
Solitary trunk feather leaf palm.

## BRONGNIARTIKENTIA

Two species, from New Caledonia.
Solitary trunk, feather leaf palm.

*Brahea decumbens*
This species has attractive blue foliage.

*Brahea dulcis*
Note the attractive blue-green leaves.

*Brahea nitida*
In this species the fan leaves are almost circular.

*Brahea salvadorensis*
This palm has attractive large fan leaves.

*Burretiokentia vieillardii*
This juvenile plant will eventually grow to 8m (25′).

## BURRETIOKENTIA

Two solitary species, natives of New Caledonia, growing in wet forests at altitudes of 50–400m (150′–1250′) on the northern and eastern sides of the island.
*B. vieillardii* (New Caledonia). Has a long, slender ringed glossy brown trunk 8–10m (25′–35′) high, topped by a crownshaft. There are about nine or ten flat, feathery leaves. The dense inflorescences are borne below the crownshaft and are followed by red fruits 15mm (3/4″) long.
**Culture and habitat** In New Caledonia, this palm grows in thick stands of tree ferns and mosses where it is cool and damp and sheltered by tall trees. Coming from this cooler elevation in a tropical area increases the adaptability of *Burretiokentia*: it will grow in temperate zones as well as tropical and subtropical. However, frosts must be avoided and in the cooler zones it will need a warm suntrap with the protection of other vegetation and top cover to keep off any cold winds. In the hotter zones it needs top cover against the hot sun. A surface mulch will also provide humidity and keep the roots cool, to simulate their natural habitat. Water often, wetting down the foliage.
**Sex** Monoecious. Self pollinating for seed production.
**Position indoors** No precise information on indoor use, but the habitat suggests that it may be suitable for indoors.
**Propagation** By seeds, but no germination times are available.

## BUTIA

Eight solitary species, located in cooler, drier areas of South America from Brazil to Argentina, often in lowland grasslands and woodlands.
*B. bonnetii* (Brazil). This species has a short thick trunk. It is a relatively small grower with small fruits.
*B. capitata*, Jelly Palm (Brazil and Uruguay and Argentina). Trunk height of this palm may vary from 2–6m (6′–20′). Old leaf bases persist near the crown of about 45 silver to blue-green recurving feathery leaves 1.5m (5′) long. The 70cm (30″) inflorescence bears orange-yellow edible fruits 3cm (1″–1 1/4″) long. This species hybridises freely with other genera, and two such crosses are included in this book. There are also many variations within the species and these follow.
*B. capitata* var. *odorata*. This tall variation can be seen in Brisbane Botanic Gardens.
*B. capitata* var. *strictior*. Origin unknown. First located specimen is growing in Huntington Gardens, California, but appears unknown elsewhere.
It has a thick trunk with clustered, upright fronds.
*B. eriospatha* (Brazil). The heavy trunk of this species is 1–3m (3′–10′) high. Leaf bases fall off sooner than those of *B. capitata*, and are covered with brown wool, as are the floral bracts. Leaves are greener. The 2cm (3/4″) fruits are yellow.

*B. yatay*, Yatay Palm (Argentina, Paraguay, Uruguay). This species has a sturdy trunk covered with old leaf bases. The spreading fronds are silvery green.
*B. yatay* var. *paraguayensis* (Paraguay). With this variety leaves are more blue-green in colour.
***Butia capitata* x *Jubaea chilensis*.** This interesting cross seen in Brisbane Botanic Gardens between two similar species, produces a thicker trunk, plus a silver colouring not found in *Jubaea*. It also inherits the cold-hardiness of Jubaea and Butia.
***Butia capitata* x *Syagrus romanzoffiana* (*Butiarecastrum nabonnandii*).** This is also a very hardy cross with the toughness of both parents carried through.
**Culture and habitat** *Butia capitata* is a hardy outdoor palm, widely grown in some areas. Its fruit can be strained to form a jelly, or fermented into a wine. Hence it is widely known as the Wine Palm or Jelly Palm. It is a fairly hardy palm, and adapts to habitats ranging from high-rainfall warm areas to drier, frosty regions. *Butias* are nice specimen trees in a mixed palm garden, making a contrast to the green colour of most other palms; they may also be planted in rows, or used as driveway liners, to good effect. Use a complete plant food at regular intervals during the growing period (warmer months). Always feed at flowering time to plump out the fruits.
**Sex** Monecious. Self pollinating for seed production.
**Position indoors** Nothing is known about *Butias* having been used indoors, perhaps because of their droopy appearance in the juvenile form. If you wish to try them, observe regularly for tip burn, then stage

***Butia capitata***
**Wine Palm or Jelly Palm.**

***Butia bonnetii***
**Note the recurving silver leaves.**

***Butia capitata* var. *odorata***
**At Brisbane Botanic Gardens.**

*Butia capitata 'Strictior'*
An unusual form from Huntington Gardens, California.

*Butia yatay*
The Yatay Palm of Argentina.

*Butia eriospatha*
The leaves are greener than those of *B. capitata*.

*Butia yatay* var. *paraguayensis*
A Paraguayan variety of the Yatay Palm.

then stage them back outdoors again, and eventually into full sunlight. Keep moist but not standing in water.

**Propagation** Grown from seeds, which are slow to germinate, taking up to 6 months.

## CALAMUS

About 370 solitary or clustered species, with eight native to Australia, and the others distributed in equatorial West Africa and from South-West and South-East Asia through to Fiji, in tropical, subtropical and monsoonal rainforests.

The Australian species are:

*C. aruensis* (Cape York area, New Guinea and Aru Islands of Indonesia). Clumping climber with stems 25m (80′) by 2cm (¾″). Leaves feathery, with leaflets 45cm (18″) by 6cm (2½″) and a prickly leaf extension to 4m (12′) long. Fruits are cream and 1cm (⅜″) long. Not pictured.

*C. australias,* Lawyer Cane, Wait-a-While (north-eastern Queensland). This vigorous climbing feather palm has numerous canes 15mm (½″) in diameter, densely armed with long red needle-like spines. The flowers in long spiky whip-like panicles are greenish and the pea-like fruits are creamy white and edible. This species grows widely on the lowlands and the Atherton Tableland.

*C. caryotoides,* Fishtail Lawyer Palm (north-eastern Queensland). A smaller and more slender species,

A *Butia* cross
*Jubaea chilensis* x *Butia capitata.*

A *Butia* cross
*Butiarecastrum nabonnandii*
(*Butia capitata* x *Syagrus romanzoffiana*)

*Calamus australis*
Hairy Mary, Lawyer Cane or Wait-a-While.

*Calamus caryotoides*
Fishtail Lawyer Palm.

*Calamus warburgii*
Attractive as a pot-plant.

*Calamus hollrungii*
This lawyer palms has a single trunk.

*Calamus muelleri*
A compact species, ideal against a tree trunk.

but still with vicious needle spines. The yellowish-green leaflets are of the fishtail type. Flowers are greenish and the edible fruits are cream to yellow, 7mm (¼″) in diameter. Excellent pot-plant.

*C. hollrungii* (coastal north-eastern Queensland and New Guinea). This species has a single spiky trunk 25m (80′) by 2.5cm (1″). Leaves are feathery with leaflets in groups along the stalk, which extends to 3m (10′) as a spike whip. The 1cm (⅜″) fruits are scaly and greenish white.

*C. moti* (coastal north-eastern Queensland). This palm resembles *C. australis*, but with distinctive clusters of heavy orange spines on the stems. Not pictured.

*C. muelleri*, Wait-a-While (coastal northern New South Wales to south-eastern Queensland). The most southerly growing species. The slender but quite long trailers are armed with spines at first, but *become less formidable* as the leaves fall and the stems become bare. The flowers are greenish and hard, and the 7mm (¼″) fruits are yellow.

*C. radicalis* (coastal north-eastern Queensland). This species resembles *C. australis*, but the long, spiked, whip-like flowering panicles are branched many times. Not pictured.

*C. warburgii* (Iron Range, north Queensland and New Guinea). A multistemmed species with stems 40m (130′) by 3cm (1¼″) it has feathery leaves with a spiky whip-like extension. Fruits are scaly, yellowish, 1cm (⅜″) long.

Other species are:

*C. filispadix* (Philippines). The juvenile specimen I saw had bright green feathery leaves with lightly spined green stalks, brownish and spinier at the base and on the trunk.

*C. reyesianus* (Philippines). This species has dark green feathery leaves with stalks covered in long brown spines, their bases brownish and more thickly spined. It suckers freely and makes an attractive pot specimen.

*C. usitatus* var. *palawanensis* (Philippines). In this variety the feathery light green leaves are longer and more narrow, and the young opening leaf ends are attached to one another by a string round the leaf. The leaf stalk is short and covered in black spines, and extends beyond the leaf with end spines that hook onto anything it touches. An attractive pot-plant.

**Culture and habitat** These savagely armed scrambing palms reach from the ground to the tops of the tallest jungle trees. On a drive up the Atherton Tableland a number of the Australian species may be observed growing up the trees beside the road. On a walk through any of the jungle tracks, you will often see the long thorny upright suckers armed with dense spines ready to attach onto the first object available to assist their growth upwards.

The canes of *Calamus* are used in the manufacture of rattan furniture. This is a sizeable and economically important industry in both the Philippines and Indonesia.

*Calamus* species, because of their growth characteristics, are best grown in botanic gardens. However, the shorter, hardier *C. muelleri*, *C. caryotoides* is

*Calamus filispadix*
A Philippine rattan.

*Calamus reyesianus*
Canes of *Calamus* species are used for making furniture in the Philippines.

*Calamus usitatus* var. *palawanensis*
Another Philippine rattan.

*Calyptrocalyx spicatus*
A single-trunked palm from the New Guinea region.

suitable for garden 'rainforests', where it will not get out of hand as the others would.

Should you wish to add a *Calamus* to your palm collection, one may be planted at the foot of a tree in a sheltered rainforest area of the garden. Warmth and daily waterings are beneficial. Keep under control by regular removal of uncontrollable canes.

It may be difficult to obtain a specimen as nurseries rarely stock *Calamus.*

**Sex** Monoecious. Self pollinating for seed production.

**Position indoors** A sunroom would be best. I have found these palms interesting and quite hardy under sunroom conditions, in a position where they receive no sun, but good light.

**Propagation** Fresh seeds germinate easily in 2–3 months. They prefer shade and plenty of water.

## CALOSPATHA

One species *C. scortechinii,* from the Malay Peninsula.

A solitary trunk climbing rattan, with feather leaves.

## CALYPTROCALYX

This solitary or clustered genus of 38 species now includes *Paralinospadix.* They grow as understorey palms in the jungles and rainforests of the Moluccas and New Guinea, growing from sea-level to mountain slopes up to 1000m (3000').

*C. spicatus* (Moluccas). This palm has a ringed trunk to 12m (40') topped with a fine head of slightly drooping feathery leaves with narrow segments. The most distinctive feature is the 2m (6') long, unbranched inflorescences, which produce orange fruits.

*C. hollrungii* (prev. *Paralinospadix)* (New Guinea). A small clustering palm for shady areas. Has numerous slender 1–1.5 cm (⅜"–⅝") branched stems topped with green or reddish-green V-shaped leaves, some with two to four leaflets. The inflorescence is a single rat-tail spike and the 15cm (½") oval fruits are red. An excellent pot-plant.

*C. microspadix* (prev. *Paralinospadix)* (New Guinea). A cluster palm with attractive red trunks to 2–4m (6'–12') or more, each topped with a graceful head of green feathery leaves on red stalks. Fruits are bright red. A handsome pot-plant.

**Culture and habitat** These small-growing, shade-loving palms grow in the understorey of tropical jungles. They closely resemble *Linospadix* , and are related to this less tropical species.

They require protection from cold winds, a humid atmosphere and regular moisture. Tropical and subtropical climates, with shade and protection from winds, would be best for outdoor plantings, the home rainforest being the ideal site. A surface mulch of leaf mould would provide the required humidity, along with daily waterings. In warm temperate areas it could also be possible to find a suntrap area that would be suitable; otherwise a glasshouse or similar protection would be required.

Do not forget that these are excellent pot-plants, and growing them this way will allow them to be moved around seasonally to maintain their cultural needs.

**Sex** Monoecious. Self pollinating for seed production.

**Position indoors** Little is known of the indoor hardiness of *Calyptrocalyx.* They do have the features of an excellent indoor plant, and are well worth a trial. Keep them out of hot sunny corners, but select a position with good natural light. Place the pot on a larger than normal saucer filled with pebbles and keep these damp for extra humidity. As regularly as is convenient give the foliage a hosing, or a shower, for moisture maintenance as well as cleanliness.

**Propagation** No times are available for germination, but it would be expected that bottom heat would be required away from the warm tropics.

*Calyptrocalyx* hollrungii (prev. *Paralinospadix*)
A small clumping palm.

## CALYPTROGYNE

Five solitary species, distributed from Mexico and Guatemala to Colombia, some growing at elevations below 700m (2000′) others from 700–2200m (2000–7000′), as understorey plants in tropical rainforests, or along streams or in swamps.

*C. ghiesbreghtiana* (Mexico). A dwarf stemless, shade-loving palm, with clumping stems and irreg-

*Calyptrogyne ghiesbreghtiana*
A showy dwarf-trunked palm.

*Calyptrocalyx microspadix* (prev. *Paralinospadix*)
A small clustering palm

*Calyptronoma rivalis*
Manac Palm. Now endangered in its habitat.

*Calyptronoma dulcis*
An attractive Cuban palm.
(Photo C. Migliaccio)

ularly divided feathery leaves. The inflorescences grow in sets of three, and flowers are followed by oval fruits 15cm (½″) long. Good as a pot-plant.
*C. sarapiquensis,* Manac Palm (Costa Rica). A stemless palm with a subterranean trunk and feathery leaves with irregular segments. The rat-tail inflorescences later bear oval fruits which are purple to black when ripe, and 15mm (½″) long.
**Culture and habitat** *Calyptrogyne* palms resemble and have been separated from *Geonoma* and *Calyptronoma.* They are stemless palms, growing in low light on the jungle floor in Central America.
*C. ghiesbreghtiana* has been cultivated as a pot-plant, and is also quite hardy in a shaded, root-free position in a tropical garden. It requires a rich soil, so dig in plenty of humus-rich additives and apply a liberal mulching of leaf mould on the surface. Water regularly to create a high humidity. Best in tropical or subtropical regions. A favourite pot-plant in glasshouses or conservatories in cooler regions.
**Sex** Monoecious. Self pollinating for seed production.
**Position indoors** No detailed information available, but well worth a try.
**Propagation** No germination times are available, but bottom heat is recommended away from the tropics.

## CALYPTRONOMA

Three solitary species, from the Greater Antilles Island, the West Indies, growing in swamps near the coast, beside streams and in wet mountainous places.
*C. dulcis* (Cuba). This palm resembles a Coconut Palm, but with a nice head of stiff deep green leaves the leaflets of which droop down on either side of the mid rib. The oval fruits are 25mm (1″) long.
*C. rivalis,* Manac Palm (Puerto Rico). The handsome robust appearance of this palm attracted my attention as I was strolling through Brisbane Botanic Gardens. Its colour and robust leaves recall a Coconut Palm. It has a trunk, without a crownshaft, to about 10m (30′) high. The broad feathery leaves are shiny, smooth and bright green. They are borne on thick golden stalks similar to those of Coconut Palms. However, this is where the similarity ends, as the oval fruits are much smaller, only 15–20mm (½″–¾″) long.

Reports indicate that this palm is an endangered species in its own country, hence seeds have been distributed around the world to safeguard its continuance. It is hardly likely to be obtainable commercially.
**Culture and habitat** *Calyptronoma* grows well in tropical and subtropical climates and does not appear to have any special requirements other than regular watering.
**Sex** Monoecious. Self pollinating for seed production.
**Position indoors** No information is available on its use.
**Propagation** Germination appears to be slow, and because it is tropical, bottom heat may be best.

## CAMPECARPUS

One species, *C. fulcitus,* from New Caledonia. Solitary trunk, feather leaf palm.

## CARPENTARIA

One solitary species, from Australia's Northern Territory, growing at low elevations in rainforests along the banks of streams and brackish estuaries.
*C. acuminata,* Carpentaria Palm (Northern Territory). This tall, slender feather palm has a smooth, ringed grey trunk 15–20m (50′–70′) high. The sparse, arching leaves are dark green and there is a short green crownshaft. The inflorescences of creamy white flowers produce red 15mm (½″) fruit, generally around December. A majestic and graceful palm. Sometimes known as *Kentia acuminata.*
**Culture and habitat** This palm is a native of northern tropical Australia particularly areas around Darwin, where it is also being used for town plantings. It seems to have adapted quite well for subtropical climates, and struggling specimens may be seen as far south as Sydney. Perhaps it would also grow in the shelter of suntraps or the protection of other vegetation in coastal frost-free areas of the warm temperate zone away from cold winds; otherwise, a glasshouse would be necessary in that zone. In the warmer subtropics and tropics, only protection from cold winds is required, along with ample water. It does not like to dry out.
**Sex** Monoecious. Self pollinating for seed production.
**Position indoors** While this palm has only had limited indoor use, reports indicate that it stands up quite well indoors provided the natural light is good. Select a warm and bright position and water well, without standing it in water. Feed occasionally during the summer growing months with a complete plant food and wet down regularly with a foliar fertiliser to maintain a lush appearance. In cooler areas, when not being grown indoors, *Carpentaria* would need a glasshouse or sheltered, warm conditions to maintain a good appearance.
**Propagation** Seeds need to be very fresh and germinate in about 2–3 months. Protect the growing plants from cold.

## CARPOXYLON

One Species, *C. macrospermum,* from the New Hebrides.

## CARYOTA

12 solitary or clustered species, occurring from South-East Asia through to northern Australia, from sea-level to 2000m (6000′) in monsoonal or humid forests.

*Carpentaria acuminata*
Carpentaria Palm.

*Caryota cumingii*
A Fishtail Palm from the Philippines.
(left)

*Caryota maxima (aequatorialis)*
A giant Fishtail Palm.

*C. cumingii* (Philippines). Similar to *C. urens,* growing to 12m (40′) with a broader trunk which shows distinct rings. A handsome specimen.

*C. maxima* (syn. *C. aequatorialis)* (Indonesia, Malaysia). A large species that has a trunk to 30m (100′) tall topped with leaves larger than those of most other varieties, growing up to 3m (10′) long. The pendulous inflorescence hangs to 2m (6′) and later carries bright pinkish-red fruits.

*C. mitis,* Clustering Fishtail Palm (India to Indonesia and the Philippines). This is the preferred species to grow because of its multiple trunks growing 5–10m (15′–30′) high in a close clump. Although each trunk dies after its final seeding, only one trunk dies at a time, leaving a nice clumped appearance. Pale cream flowers are followed by orange to red fruits about 1–2cm (⅜″–¾″) long.

*C. ochlandra,* Chinese Fishtail Palm (China). This species has a single trunk to 8m (25′) tall, with leaves that are narrower and blunter than in other varieties. It tolerates cooler conditions, including light frost.

*C. rumphiana,* Fishtail Palm (Indonesia, New Guinea, northern Australia). Similar to *C. urens* but with larger leaflets, to 45cm (1¾″). This native palm has a grey ringed trunk to 20m (60′) high.

*C. urens,* Solitary Fishtail Palm (India, Malay Peninsular). A handsome palm, with a broader trunk 30–40 cm (12″–15″) wide and about 12m (40′) high. There is a dense crown of arching leaves with narrow leaflets. The inflorescence is large, to 3m (10′) with

*Caryota mitis*
Clustering Fishtail Palm.

*Caryota ochlandra*
Chinese Fishtail Palm.

greenish flowers followed by red fruits, similar to *C. mitis.*

**Culture and habitat** *Caryotas* are most ornate and interesting palms. Their unusual leaf form, which gives them their common name of fishtail palms, has a mixed public reaction. It is either loved or hated but most people love it. When viewed against the light, the divided leaves make the most interesting shapes and patterns, distinctly different from those of most other types of palms. When seen growing amongst other trees and shrubs, a fishtail palm can easily be mistaken for some other type of tree because of the small leaf segments.

In tropical to warm temperate areas (no frost), these palms may be successfully grown in the garden. They become quite handsome with their thick trunks and interesting leaf patterns. *C. mitis* with its multiple trunks also forms an eye-catching clump. Protect from cold winds in the cooler areas.

Another very interesting feature of these palms is that they begin flowering and seeding in the upper leaf axils first, gradually reflowering and seeding down the trunk. When a trunk has set its last bunch of seeds at its lowest leaf axil, it dies. Sadly, in the case of *C. urens* and other single-trunked species, this means the death of the palm. The whole flowering process usually takes up to 7 years.

Apart from their ornamental value, *Caryotas* are valuable as sources of sugar and sago. The sap is also fermented into alcoholic drinks.

**Sex** Monoecious. Self pollinating for seed production.

**Position indoors** Smaller plants are nice in homes with advantageous natural lighting. However, larger *caryotas,* although spectacular, reach above head height, and could only be used for high-ceilinged buildings. Natural light wells help to prolong the plant's life. In cooler climates, *Caryotas* must be grown in glasshouses. Water very well but do not leave standing in saucers of water.

**Propagation** Seeds usually germinate in 3–4 months. Care should be taken in handling the fruits as they contain an irritant. Protect the young plants from cold.

## CATOBLASTUS

17 species from Panama, Colombia, Ecuador, Brazil, Venezuela and Peru.

Solitary or clustered, feather leaf palms.

## CERATOLOBUS

Six species from the Malay Peninsula and western Indonesia.

Clustered, climbing rattans, with feather leaves.

*Caryota rumphiana*
Australia's Native Fishtail Palm.

*Caryota urens*
Solitary Fishtail Palm. Has narrow leaflets.

*Ceroxylon alpinum*
Andean Wax Palm is a tall mountain-growing palm.

# CEROXYLON

20 or so solitary species from the mountains of north-western South America, from Colombia to Bolivia.

*C. alpinum* (syn. *C. andicolum),* Andean Wax Palm (Colombia and Venezuela). A tall grower with a 30m (100′) trunk, 30–60cm (12″–24″) thick and usually swollen in the middle. The crown has about 10 strong feather leaves, dark green above and powdery white beneath. The branched flowering stems hang from amongst the leaves and later bear 12mm (½″) round orange fruits, black at maturity.

*C. hexandrum* (Colombia). This tall wax palm withstands the cool conditions of high altitudes in the Colombian Andes. The leaf stalks and the undersides of the leaf blades are silver from their wax coating.

*C. quindiuense,* Andean Wax Palm (Colombia). Similar to and often confused with *C. alpinum,* but growing much taller (to 60m (200′)). The undersides of the leaves and their stalks are silver waxy.

**Culture and habitat** *Ceroxylon* species grow high (to 4000m, 12000′) in the Andes. Closely related to Australian *Oraniopsis* and to *Ravenea* of Madagascar, these attractive, eye-catching palms from elevated tropical regions grow well in the Royal Botanic Gardens, Sydney, in a warm temperate climate. Because of their fairly high-altitude habitat in their native tropics, they should grow in cooler to elevated tropical, subtropical and temperate climates that do not suffer frosts or cold winds. Being tall growers,

*Ceroxylon hexandrum*
Another attractive wax palm.

*Ceroxylon quindiuense*
Another Andean Wax Palm.

they will rise above the protection of lower vegetation. Water regularly.
**Sex** Dioecious. Self pollinating for seed production.
**Position indoors** No knowledge of indoor use.
**Propagation** Fresh seeds may be difficult to obtain, and would require hot humid conditions to germinate.

## CHAMAEDOREA

100 solitary or clustered species, distributed from Mexico to Brazil and Bolivia, growing as understorey plants in damp mixed forests, even mountain forests.
*C. atrovirens* (Mexico). The interest in this species, which grows to about 2m (6′), is in the fact that the black trunk becomes multiheaded, dividing a little above soil level. Flowers are followed by 6mm (¼″) black fruits. There are some reports that this species does not perform well indoors. My own experience with this plant indoors, was that when I allowed it to dry right out between waterings, there was some slight tip burn; but since I have kept it more moist by more regular waterings, it has thrived. My plant is positioned near a window.
*C. brachypoda*(Guatemala). Often confused with *C. stolonifera,* this species grows to 2m (6′) tall. The very slender trunks sucker out rapidly into a thick clump, and have green indented V-shaped leaves.
*C. cataractarum* (Mexico). This palm is very similar to *C. atrovirens.* The branching green inflorescence

*Chamaedorea atrovirens*
The trunks of this species divide.

*Chamaedorea brachypoda*
A Fishtail Parlour Palm.

*Chamaedorea cataractarum*
This species resembles *C. atrovirens.*

*Chamaedorea concolor*
The tall trunk of this species grows to 2m (6′).

*Chamaedorea costaricana*
Bamboo Palm forms a clump.

bears yellow flowers, followed by small reddish fruits 4mm x 6mm (about ¼″) long.

*C. concolor* (Mexico). This species has a slender, bamboo-like trunk 2m (6′), and shiny green leaves. It requires shady but warm conditions. Flowers are orange and fruits red.

*C. costaricana*, Bamboo Palm (Costa Rica). A densely clustering palm to about 3m (10′) eventually thickening to a clump about 2m (6′) wide. It is similar to *C. seifrizii*, but larger overall, and has yellow flowers and bright orange fruits. It is hardy and a strong grower.

*C. elatior*, Climbing *Chamaedorea* (Mexico). A scrambling species with a weak floppy trunk to 4m (12′), this palm needs support and does best in shade. The small black fruits are borne on orange to brownish stems amongst the 50–60cm (20″–24″) green leaves. It is best multiple-planted.

*C. elegans*, Parlour Palm (Mexico and Guatemala). Also sold as *Neanthe bella* or *Collinia elegans*, this is the most widely grown of the species. It may be sold as a miniature, in various containers or mixed gardens. However, with continual repotting it will eventually reach 1–2m (3′–6′). It will not tolerate sun, but is very hardy indoors or in sheltered outdoor areas. While singly planted specimens are dainty as miniature pot-plants, clusters of three plants are preferable in larger pots for that clumped effect. It has yellow flowers and black fruits.

*C. ernesti-augusti* (Mexico to Honduras). This species, often confused with *C. geonomiformis* has a slender solitary trunk 1m (3′) tall. The leaves are a simple

*Chamaeedorea elatior*
Climbing *Chamaedorea*. Needs support.

*Chamaedorea elegans*
The very popular Parlour Palm.

*Chamaedorea elegans*
This is the narrow-leaved form.

*Chamaedorea ernesti-augusti*
The leaf of this species is undivided.

*Chamaedorea erumpens*
Bamboo Palm.

*Chamaedorea geonomiformis*
This species (photographed in Australia) more resembles *C. ernesti-augusti*. M flower spikes.

V shape, bright green and larger than that variety. The inflorescences are red, the female one having a single stem and the male one a branched stem; fruits are black.

*C. erumpens,* Bamboo Palm (Honduras, Guatemala). This is a densely clustering species to 3m (10′) tall by 1m (3′) wide. It only carries about 20 leaflets per frond, with the terminal pair very wide. Fruits are black.

*C. geonomiformis* (Honduras, Belize). This palm has a slender trunk 1–1.3m (3′–4′) high, with deeply indented, broad two-segmented, dark green leaves. Flowers are cream and fruits reddish. Resembles *C. ernesti-augusti* and often confused.

*C. glaucifolia* (Mexico). This species has a tall slender trunk to 6m (20′) high. The almost 2m (6′) long leaves have narrow segments coloured powdery grey over dark green. The fruits are green. For warm climates.

*C. graminifolia* (Guatemala). Reported to be two varieties with this name, one a solitary trunk and the other pictured here, on multiple trunks.

*C. klotzschiana* (Mexico). A solitary-stemmed species to 2.5m tall (8′) with the glossy green leaflets arranged in groups of five or six along the central stalk.

*C. metallica* (Mexico). A dwarf and slow-growing single-trunked species to 1m (3′). The leaves may be whole or divided into large segments and are metallic blue-green, with grooved markings. Flowers are inconspicuous and fruits black.

*C. microspadix,* Bamboo Palm (Mexico). An open-clumping species suitable for shaded, warm outdoor areas or indoors. It grows to 1m (3′) tall and has dull

*Chamaedorea glaucifolia*
The fruiting stems are bright red.

*Chamaedorea klotzschiana*
The leaflets are in groups.

*Chamaedorea metallica*
The leaf segments have divided on this specimen.

*Chamaedorea graminifolia*
This specimen has multiple trunks.

*Chamaedorea microspadix*
Bamboo Palm.

*Chamaedorea monostachys*
A rarely seen species.
(Photo C. Migliaccio)

green leaves. Flowers are creamy and fruits orange to red.
*C. monostachys* (Mexico). An unusual species resembling *C. elegans* growing in the Rare Plants House at Fairchild Gardens in Florida.
*C. neurochlamys* (Mexico, Guatemala, Honduras). This species grows taller (about 2–3m (7′–10′)) and has a thicker 5cm (2″), green, bamboo-like trunk topped with bright green feather fronds about 1m (3′) long.
*C. oblongata* (Mexico to Nicaragua). This palm has a slender, ringed trunk to 3m (10′). The shiny green leaflets are broader than in most *Chamaedorea* species. Flowers are bright orange and fruits deep green.
*C. pinnatifrons* (Colombia, Venezuela). This palm has a slim bamboo-like trunk to 2m (6′) tall. Leaves are 60cm (2′) long with yellowish green stalks. Leaflets are mid-green and broader in the centre. The inflorescence is 60cm (2′) long and has a yellow stem.
*C. pochutlensis* (Mexico). This is a tall (4m (13′)) clumping species. The long trunks resemble sticks of green bamboo topped with green feather leaves 1m (3′) in length.
*C. radicalis* (Mexico). This palm clusters from the base to a height of 3m (10′), with dark green leaves 1m (3′) long. It flowers when quite young, with the inflorescence held high above the foliage. It has orange-yellow flowers and orange fruits, and grows best in warm, shady outdoor areas.
*C. schippii* (Belize). This is a tall clumping species somewhat resembling *C. pochutlensis* in appearance and habit.

*Chamaedorea neurochlamys*
Note the heavier, bamboo-like trunk.

*Chamaedorea oblongata*
This *Chamaedorea* has broader, pouch-like leaves.

*Chamaedorea wendlandi*
This specimen was growing in a sunny situation.

*Chamaedorea pinnatifrons*
The fruiting stem is yellow.

*Chamaedorea pochutlensis*
This species grows in a tall cluster.

*Chamaedorea radicalis*
The inflorescences are unusual.

*Chamaedorea schippii*
A strong, heavy grower.

*Chamaedorea seifrizii*
Reed Palm or Clustered Parlour Palm.

*Chamaedorea stolonifera* (juvenile)
Climbing Fishtail Palm. Often confused with *C. brachypoda.*

*Chamaedorea tepejilote*
Grows on a sturdy trunk.

*Chamaedorea tuerckheimii*
The leaflets of this species are irregular.

*C. seifrizii,* Reed Palm, Clustered Parlour Palm (Mexico). Another multiple-trunked variety, with the straight upright trunks growing densely to about 3m (10′). The leaves are much the same shape as in *C. elegans.* I have grown this species in my home in quite darkish positions for 3–4 years, without ever spelling it outside. It has bright orange inflorescences and black fruits.

*C. stolonifera,* Climbing Fishtail Palm (Mexico). This species has many slender trunks to 2m (6′) tall. The leaves are deeply indented and in broad segments.

It will grow in sunshine. Flowers are orange and fruits black.

*C. tepejilote* (Mexico, Central America, north-west South America). This palm has a solitary, bamboo-like trunk to 3m (10′) tall, topped by a small crownshaft and a sparse crown of deep green leaves over 1m (3′) long. The branched, bright yellow inflorescence is followed by blue-black fruits.

*C. tuerckheimii* (Belize, Guatemala, Mexico). This species has a greyish solitary trunk. It differs from others in that the leaflets resemble those a *Caryota,* being broad and triangular.

*C. wendlandi* (Mexico). This solitary trunk variety is being tried in Brisbane Botanic Gardens in full sun.

*C. woodsoniana* (Panama, Costa Rica). The slender trunks of this species grow to about 2m (6′) and the feathered fronds, 60cm (24″) long, have deep green tapered leaflets. It needs shade.

*C.* sp. 9087, bed 25, at Royal Botanic Gardens, Sydney. This tall species caught my attention with its unusually thick trunk, along with the crownshaft and the long (90 cm (3′)) inflorescence stem. There was also another species, labelled *C.* sp. 780622, seed 1978, which appeared much the same.

**Culture and habitat** *Chamaedoreas* are popularly grown as indoor pot-plants. They may also be grown outdoors as undercover palms in warm frost-free areas. Because of their need for shading from the sun, most varieties yellow, even in partial sunlight. With these features, coupled with the fact that they are mostly low growers, their success and popularity indoors, or outdoors as shaded garden or pot-plants is assured.

The range of *Chamaedorea* species is large and varied: they range from small single-trunked palms to taller multistemmed plants. Most of the popularly grown varieties are excellent indoors, but with over 100 known species, there are many new forms being introduced to cultivation. *Chamaedoreas* produce inflorescences when quite young. These are stems which may be coloured carrying what appear to be red or yellow small "berries", (for the male plant). These fall after a few weeks and carry the pollen. With the female, those stems have small round indents and if pollinated the young green fruits form in these. Matured fruits are generally black and about 1–2cm (⅜″–¾″) in length and oval in shape.

**In the garden** Grow only in warm positions, in total shade. Sheltered and shaded verandas, poolsides and courtyards would be best. Some varieties such as *C. seifrizii* will tolerate more sun, but still yellow if they receive too much.

*Chamaedorea woodsoniana*
Note the multiple trunks.

*Chamaedorea* sp. 9087
A tall species at Sydney Botanical Gardens.

*Chamaerops humilis*
Mediterranean Fan Palm.

**Sex** Dioecious. Need a male and female for seed production.
**Position indoors** These are among the hardiest of all indoor palms. They not only tolerate the worst of indoor positions for varying periods, but also stand up to the abuse of irregular watering. These abuses should be discouraged, however, as best results are obtained where good lighting and regular waterings are maintained. Do not leave water in the saucer for more than 1 hour. Fertilise occasionally during the growing period (warmer months), using a complete fertiliser. Excess urea can cause adverse effects.
**Propagation** Seeds are best fresh, but even then germination can be irregular, taking 1–6 months. I have known two growers sharing the same seed batch to have opposite results. As an alternative to raising from seed, some of the multistemmed varieties may also be divided, when sufficiently clustered.
**Pests and problems** Because they are mostly used indoors, *Chamaedoreas* have few caterpillar problems, but if grown outdoors they should be regularly observed. They do suffer from mite attack.

## CHAMAEROPS

One clustering species from both the African and European fringes of the Mediterranean Sea. It can be found growing from the sandy seaside fringes or rocky hills to 1000m (3000′). In the wild they are trunkless, but they develop trunks in cultivation.

*Chamaerops humilis* var. *elatior*
A variety growing at Fairchild Gardens, Florida, USA, see notes.

*Chamaerops humilis*
Ideal as an attractive pot palm

*C. humilis,* Mediterranean Fan Palm (Mediterranean fringes). A widely grown clustering fan palm which is extremely hardy, even growing to altitudes of 1000m (3000′) and tolerating moderate snowfalls. Its form may vary from a single trunk to 6m (20′) to clumps of ground-covering trunks of only 1–2m (3′–6′). These trunks are usually covered with old, spent leaves, which may be trimmed off to leave the old leaf bases. The leaves are a fan of narrow, but stiff, deep green to grey-green segments, and up to 1m (3′) in size. The inflorescence is short and hidden amongst the leaves; flowers are yellow and fruits are brown to yellow, varying in size but generally around 2 cm (¾″) long.
*C. humilis* var. *elatior.* This variety is growing at Fairchild Gardens, Florida, and the name was listed by Dr. Moore as synonymous with the old name, *C. macrocarpa.*
**Culture and habitat** *Chamaerops* is a native of the areas of southern Europe and northern Africa surrounding the Mediterranean Sea. It is widely grown all round the world and can be considered hardy, having adapted well to most climates. It is suitable for all zones and will grow anywhere from seashores to low mountains, in shade or hot sun. Once established it will withstand dry conditions. It can be grown as a patio or poolside pot-plant. Plants are usually low-growing and spreading, so give them adequate room when planting out.
**Sex** Dioecious, occasionally hermaphrodite. May produce a few seeds but needs a male and female for seed production.

**Position indoors** My only trial resulted in tip burn, and I did not persevere. However, I notice they are highly recommended by others, so they are worth a try.
**Propagation** Propagation is from seeds, which germinate in 3–4 months. In the cold, growth is slow. It should also be possible to grow *Chamaerops* by rooted division from an established clump.

## CHAMBEYRONIA

Two solitary species from New Caledonia, one *(C. macrocarpa)* occurring in wet forests, the other *(C. lepidota)* on shale soils only.
*C. macrocarpa* (syn. *C. hookeri).* A very striking palm, as the new leaves open bright red and are very eye-catching. The 20m (60′) ringed trunk has a green crownshaft and the green feather leaves, paler beneath, curve downwards. The inflorescence is short and branched and later bears 2cm (¾″) red fruits.
**Culture and habitat** *Chambeyronias* are slow-growing handsome palms closely resembling *Archontophoenix.* They come from forested areas up to 1000m (3000′) altitude with high rainfall. Although they originate from tropical New Caledonia, they appear to grow quite well in climates as far south as Sydney, with fine specimens growing in the Royal Botanic Gardens there, as well as in Foster Gardens, Hawaii.

*Chambeyronia macrocarpa*
A very dainty grower.

*Chambeyronia macrocarpa*
This plant is displaying a new red leaf.

*Chelyocarpus chuco*
An ancient palm species.
(Photo C. Migliacco)

*Chelyocarpus ulei*
The leaves are in the form of a lovely segmented circle.

*Chamberyronia* is a handsome palm outdoors and trials may prove it to be as hardy indoors. Unfortunately scarcity of stock lessens the chances of finding out more cultural information.

Although I had seen *Chambeyronias* before, I had never seen them in all their glory, with their eye-catching red leaves, until my last trip to Hawaii. This colouring makes them magnificent amongst palms. They may be grown from tropical to warm temperate climates, with protection from cold winds in the latter. This may be best achieved by planting them amongst trees for protection. They should be watered well, and ground mulches will help provide humidity. They are slow-growing.

**Sex** Monoecious. Self pollinating for seed production.

**Position indoors** No information is available on their use indoors, perhaps because stocks are so rarely available.

**Propagation** No information is available. As they are closely related to *Archontophoenix*, perhaps similar treatment would work.

## CHELYOCARPUS

Three solitary or clustered species from the Amazon area of Bolivia, Brazil, Peru and West Colombia, occurring in high-rainfall areas at low elevations.

*C. chuco* (Brazil). This is an unarmed medium-sized cluster palm to 7m (23′) with slender trunks on which the old leaf bases persist. The broad fan leaves are split toward the edges into wedge-shaped sections, as in *Licuala*. The branching inflorescence later develops corky-surfaced roundish fruits, 25mm (1″) in diameter.

*C. ulei* (Central America). In this species the leaves are more deeply divided than in *C. chuco.*

**Culture and habitat** *Chelyocarpus* species are ancient palms and most resemble *Thrinax* and *Trithrinax* in botany and form. They originate from tropical high-rainfall areas, and this consideration should be remembered in their culture. They appear to grow well within the tropics. If you intend to grow them in cooler latitudes than that, care should be taken to select a warm suntrap with top cover and ample, regular watering.

**Sex** Hermaphrodite. Self pollinating for seed production.

**Position indoors** No information is available.

**Propagation** No information on germination is available. However, as *Thrinax* and *Trithrinax* are both closely related, the same techniques could be tried.

## CHRYSALIDOCARPUS

20 solitary or clustered species from Madagascar and Nossi Be Island. Growing in rainforests or along the banksland of watercourses, from sea-level to medium altitudes. *C. lutescens* occurs in sand dunes around watercourses.

*C. cabadae,* Cabada Palm (Madagascar). This palm

has clustering trunks to about 9m (30′) tall which are green ringed with grey. The feathery leaves are deep green. The flowers are creamy yellow and the 15mm (½″) fruits red. It is fast-growing.

*C. lucubensis* (Nossi Be Island). This species has a slender, cane-like, ringed grey trunk to 10m (30′) high. The short crownshaft at the top is lime green and the fronds, instead of being flat, have the leaflets in various planes, giving a plumose appearance. The fruits are 15mm (½″) long, oblong and black. Can be used indoors where height permits.

*C. lutescens,* (Madagascar) Golden Cane Palm, Butterfly Palm. A multitrunked palm, ranging by degrees from a taller plant with one trunk to a short, heavily clustered plant. If you are multiple-planting, it is often best to vary the sizes and types so that you get both height and thickness. When planted in the garden, the more clustered plants form a thicker cluster with more trunks, while the less clustered plants have fewer but heavier trunks. On my last visit to Singapore, I observed that this palm had replaced all other indoor palms throughout the hotels and offices I saw. The fruit is violet-black and oblong, 16mm (⅝″) long.

*C. madagascariensis* (Madagascar). This species, with two or more heavier grey trunks, is similar in appearance to *C. lucubensis* (which is said to be a single-trunked form). The short crownshaft is covered in grey scurf. The branching inflorescence has yellow flowers and fruits are yellowish and 16mm (⅝″) long.

**Culture and habitat** Wrongly sold as *Areca* in many countries, these very dainty palms are a gem in warm

*Chrysalidocarpus cabadae*
Cabada Palm.

*Chrysalidocarpus madagascariensis*
This species has two or three thick trunks.

*Chrysalidocarpus lucubensis*
An attractive single-trunked species.

*Chrysalidocarpus lutescens*
Golden Cane or Butterfly Palm.

*Chrysalidocarpus lutescens*
Seen at its best in a large clump.

climates as garden specimens, as well as being extensively used as indoor palms. Of the 20 species, *C. lutescens* is the most widely grown for these purposes. In cool climates they would need to be grown in glasshouses for best results. However, as outdoor specimens are growing in botanic gardens in both tropical and temperate zones around the world, it may be possible to grow these palms in warm suntrap positions in those cooler zones.

**In the garden** Best in warm climates. They should also grow in selected frost-free, sheltered areas and suntraps in warm temperate areas. For best results, some potting compost should be dug in before planting. Hose regularly to retain humidity.

**Sex** Monoecious, except for *C. lutescens* which may be dioecious. Mostly self pollinating for seed production.

**Position indoors** *C. lutescens* is widely used indoors in many areas of the world, and here I refer mainly to that species. Select a bright, warm position and keep well watered. Winter warming is necessary in cooler climates. Multiple plantings look good and will give you a clustered effect more quickly.

**Propagation** Seeds of *C. lutescens* take 1–2 months to germinate, while those of *C. lucubensis* take 2 months or more and require early shade; *C. madagascariensis* appears to take up to 5 months. Clumps of *C. lutescens* may be divided.

*Chuniophoenix humilis*
A little-known palm resembling *Rhapis.*

# CHUNIOPHOENIX

Three clustered species from Vietnam, southern China and the island of Hainan. Located growing in the understorey of forests.
*C. humilis* (Vietnam). A delightful small palm much resembling *Rhapis* in appearance. It has multiple ringed trunks that appear to lack the hairs of *Rhapis* trunks. The leaves are similar in shape to those of *R. excelsa,* but are mid to bright green. Inflorescences are borne amongst the leaves. The red fruits are roundish and 2cm (¾″) in diameter.
**Culture and habitat** Little is yet known about this 'new' palm in cultivation. The specimen photographed was growing out of doors and under shade. It, and three or four other specimens, were thriving in a warm subtropical climate without any obvious problems.
**Sex** Hermaphrodite. Self pollinating for seed production.
**Position indoors** Not known.
**Propagation** Little is known about germination of seeds or division of clumps.

# CLINOSPERMA

One species, *C. bracteale,* from New Caledonia. Solitary trunk, feather leaf palm.

# CLINOSTIGMA

13 solitary species native to a number of Pacific Island groups between New Guinea and Fiji, growing in rainforests along ridges and crests. Their timber is used in home building.
*C. ponapense* (Ponape Island, Carolines). A tall, handsome palm to 12–15m (40′–50′), with a smooth ringed grey trunk topped with close-feathered fronds. The inflorescences are short and branches, and the 2–3 cm (¾″–1¼″) fruits are blackish.
*C. samoensis* (Samoa). A handsome feather palm with a ringed, grey trunk. It has about eight to 10 deep green fronds above a long green crownshaft. The inflorescences are much branched, and the 5mm (2″) fruits are red.

**Culture and habitat** These rare and beautiful palms are little grown at present. Their natural habitat is tropical, about 10° N, with an abundant daily rainfall. The lovely specimen of *C. ponapense* photographed was growing at Foster Gardens, Hawaii, where it has grown to maturity and is now seeding.

**Sex** Monoecious. Self pollinating for seed production.

**Position indoors** No information is available.

**Propagation** No information is available, other than germination after 48 days. However, as these palms are tropical, germination may be more even with bottom heat.

*Clinostigma ponapense*
A tall grower

*Clinostigma samoense*
A palm from Samoa.

*Coccothrinax alta*
The leaves are silver beneath.

# COCCOTHRINAX

49 solitary or clustered species from the West Indies, with the greater number in Cuba. They are distributed from the coast to inland valleys of the highlands, growing in dry limestone or serpentine rocky areas.
*C. alta* (Puerto Rico). A medium-sized, slow-growing palm with a trunk that is bare below, but matted with fibres above, to 8m (25′) tall. The graceful, drooping green fan leaves are silver beneath. Flowers are white and the 1cm (⅜″) fruits are brownish to black.
*C. argentata,* Florida Silver Palm (Florida). The trunk may reach to 6m (20′) and carries about 15 leaves about 1m (3′) wide. These are droopier and more silver beneath than in other species.
*C. argentea,* Hispaniolan Silver Palm (West Indies). The trunk of this palm reaches up to 10m (30′). The fan leaves are heavily segmented, dull green above and silvery beneath.
*C. barbandensis* (as seen in Fairchild Gardens). A tall, usually solitary palm, with a root boss and a slender trunk to 10m (30′). The trunk is topped with a loose head of green fan leaves with 1m (3′) green stalks.
*C. crinita* (Cuba). An oddity because the trunk, even when quite small, is fully covered in a mass of brown woolly hair. It grows around 5m (16′) tall and is topped with deeply segmented round green fan leaves, which are greyish beneath. It is sun-tolerant early. Flower sprays hang as low as the leaves: fruits are purplish black and 1 cm (⅜″) long.

*Coccothrinax argentata*
Florida Silver Palm in its natural habitat.
(Photo C. Migliaccio)

*Coccothrinax argentata*
Florida Silver Palm.

*C. dussiana* (Guadeloupe). This species has a slender trunk to 10m (30′) high topped with about 18 large, almost circular deep green fan leaves which are slightly whitish beneath. Fruits are 8mm (3/8″) long and black. This may be the same species as *C. barbadensis, C. dussiana* being an old name.
*C. ekmannii* (syn. *Haitiella ekmannii* ), Gouane Palm (Hispaniola). This is a small erect fan palm, which may grow differently under different conditions. The specimen photographed was growing in a garden. In its natural habitat it grows in rock and coral crevices. The trunk, up to 9m (30′) tall, is extremely slender, but will withstand a hurricane. The stiff fan leaves are silver grey. The inflorescences are small and branched, and the fruits are very small.
*C. fragrans* (Cuba). The grey trunk of this species reached 6m (20′) and carries about 12 attractive fan leaves similar to those of *C. argentea.* The yellow flowers are fragrant.
*C. miraguama* (Cuba). The trunk of this species is bare below and covered with matted brown fibres above. The head of fan leaves is most attractive, the segments of each leaf forming a beautiful circle. The flowers are yellow and the fruits, 1cm (3/8″) long, are first red, then black.
**Culture and habitat** These fan palms from the Caribbean are becoming more widely grown in Australia. As they are of moderate size and quite hardy, they should appeal. They are somewhat similar to *Thrinax* but differ in such features as the seed colour and the silvery foliage.
In tropical and warm temperate areas *Coccothrinax*

*Coccothrinax barbadensis*
Planted as a clump, shown in the centre of the picture.

*Coccothrinax argentea*
Hispaniolan Silver Palm.

*Coccothrinax crinita*
Note the thick wool on the trunk.

*Coccothrinax dussiana*
This is an old name for *C. barbadensis.*

*Coccothrinax ekmannii*
Gouane Palm syn. *Haitiella ekmannii*

*Coccothrinax fragrans*
The yellow flowers are scented.

*Coccothrinax miraguama*
The foliage is distinctive.

will grow in an open garden, but in cooler regions they would require suntrap protection. They are not used indoors, but do make wonderful potted palms out of doors. They may be grown in full sun, or under the shade of other plants. They show some tolerance of poor drainage, but they prefer limestone soils with good drainage. They may be used for seaside plantings, as they have good salt and wind tolerance. Attend to watering during dry periods.

**Sex** Hermaphrodite. Self pollinating for seed production.

**Position indoors** No indication of use indoors.

**Propagation** Fresh seeds germinate within 1–2 months, but growth is slow. A few varieties take longer. They require a soil rich in lime (alkaline pH).

## COCOS

One solitary species, occurring around all tropical coastlines. Now thought to have originated from the western Pacific, possibly Australia or the Philippines.

*C. nucifera,* Coconut Palm (Pacific and Indian Ocean islands). This graceful palm with a leaning trunk growing to 25m (80′) is one of the world's most valuable commodities. Everyone knows its importance to the economies of many countries. In the tropics there are many strains of coconuts, selected for size of plant as well as for fruit features. These strains include 'Dwarf Golden Malay', 'Dwarf Green', 'Dwarf Samoan' and many others in fact every locality boasts a superior strain.

**Culture and habitat** This well-known palm grows on the coastlines of tropical islands. Coconut Palms will also adapt to more inland habitats, and in many tropical areas can be seen firmly established both inland and on mountains.

Almost every part of the Coconut Palm is used, the trunks as timber, the leaves for thatching, weaving and wrapping food for cooking. From the fruit we get coconut flesh and also oils used in many ways. Even the husk has an everyday use. The milk is drunk and an alcoholic drink is made by tapping the sap from the flower stalk.

Coconuts may be grown, and cropped in tropical and subtropical climates, however, cropping is doubtful in that cooler region.

They will tolerate salt water and sea winds, and so can be valuable along the coastline, and along brackish rivers and canals. They need a hot position, in full sunlight.

**Sex** Monoecious. Self pollinating for seed production.

**Position indoors** Little used, but I daresay they could be useful in glasshouses, sunrooms and conservatories. I have seen a 'shot' Coconut Palm potted up and used indoors in a shop. Upon asking, I was told that it had been there for about a year. I was impressed because there were no signs of distress at all. I also observed that the soil appeared quite dry. Whether by intention, or just neglect, I do not know.

**Propagation** Propagated by planting coconuts half-

*Cocos nucifera*
Coconut Palm.

*Cocos nucifera 'Malay Dwarf'*
One of the dwarf cultivars.

*Copernicia alba*
There are still large stands of this palm in nature.

buried in soil. They germinate in 4–5 months. Allow to root before potting up.
**Pests and problems** In cooler climates there is a high fatality rate in getting young coconut trees away, usually 50 per cent. So do not be discouraged if at first you do not succeed. Spring plantings would be best, to allow a maximum time for acclimatisaton.

## COPERNICIA

25 solitary species, mainly from Cuba; two from Hispaniola and three from South America. The West Indian species are from dry wooded lowlands, and the Brazilian ones from areas subject to flooding.
*C. alba* (South America). A sturdy, tough grower, and plentiful in the wild, this species has a strong trunk with a large head of fan leaves, much resembling *C. prunifera.*
*C. baileyana* (Cuba). The outstanding feature of this palm is its solid trunk to 12m (40′) topped off with a fine array of green fan fronds with waxy white spiky stalks. The flower stems arch down well below the leaves. The 2cm (¾″) fruits are dark brown.
*C. cowellii* (Cuba). This species has a sturdy patterned trunk and is rather similar to *C. rigida* and *C. macroglossa,* but differs in that its leaves have wider segments and the dead skirt has a much coarser appearance.
*C. glabrescens,* Guano Palm (Cuba). This palm has single or multiple suckering trunks to 5m (15′) high

*Copernicia baileyana*
Note the hanging inflorescences.

*Copernicia cowellii*
Note the inflorescences.
(Photo C. Migliaccio)

topped with stiff deep green fan leaves with short stalks. The inflorescences are held above the leaves and the brown fruits are about 12–15mm (½″) long.

*C. hospita* (Cuba). Similar to *C. rigida;* a very sturdy and rigid grower, with stiff blue-green fan leaves which form a thick dense skirt as the palm slowly grows on its stout 12–15m (40′–50′) trunk.

*C. macroglossa* (syn. *C. torreana),* Cuban Petticoat Palm (Cuba). Grows to about 10m (30′) and unless the mass of dead leaves is removed the trunk is invisible. The central glossy green fan leaves are upright, while the older ones are displayed in a layer over the dead leaves. The inflorescences spray down lower than the leaves.

*C. prunifera* (syn. *C. cerifera),* Carnauba Wax Palm (Brazil). This palm has a patterned trunk to 10m (30′) high, often carrying old leaf bases. The fan leaves have long stalks, and are commercially harvested for their wax, a valuable industry in Brazil. Branched inflorescences 2–2.5m (6′–8′) long extend from among the lower leaves, and fruits are oval, about 24mm (1″) long.

*C. rigida,* Jata Palm (Cuba). The 10–12m (30′–40′) trunk of this species becomes bare and ringed as it ages. The large stiff fan leaves have no stalk; upright at first, they gradually droop as they age. The inflorescences hang out from the leaves, and the green fruits are 12–15mm (½″) long.

*C. yarey* var. *robusta* (Cuba). The 6–8m (20′–30′) trunk of this palm is topped with a head of fan leaves with short stalks. The leaf segments split to halfway and

*Copernicia glabrescens*
A cluster of the Guano Palm from Cuba.

*Copernicia hospita*
A showy fan palm from Cuba.

*Copernicia macroglossa*
Cuban Petticoat Palm.

*Copernicia prunifera* (syn. *C. cerifera*)
Carnauba Wax Palm.

*Copernicia rigida*
Jata Palm. The cascading inflorescences are prominent.

droop at the tips. The inflorescences spray outwards from amongst the leaves.
**Culture and habitat** *Copernicias* are tropical and amongst the most spectacular palms on earth. They are fan palms, quite thorny in some cases, with a leaf arrangement that attracts attention. Unfortunately they are rarely seen away from their natural habitat, except in botanic gardens. They seem to prefer a tropical climate, but specimens are growing outdoors in subtropical Brisbane (28° S) and under glass in warm temperate Sydney (34° S).
**Sex** Hermaphrodite. Self pollinating for seed production.
**Position indoors** Not known to be used indoors.
**Propagation** Fresh seeds begin germinating within 2 months, but growth is slow.

## COLPOTHRINAX

Two species, from Cuba and Panama to Guatemala. Solitary trunks, with centre bulge and fan leaves.

## CORYPHA

Eight solitary species, occurring from Bengal (East India) and Sri Lanka and through South-East Asia to northern Australia. They grow, not in rainforests, but on open plains and around habitation, even in areas

*Copernicia yarey* var. *robusta*.
One of the smaller *Copernicia* varieties.

subject to flooding, or near watercourses. *Coryphas* do not flower until they are mature, usually 20–25 years, even longer. They then send up a magnificent head of flowers, which could produce up to a million seeds. When these have matured the plants slowly die, usually over a year.

*C. taliera.* (Bengal, East India). Now a rare palm and magnificent when in flower. The trunk is stout, less than 10m (30′) high, and untidy with old leaf bases. The stiff green fan leaves have brownish stalks. The tall branching inflorescence towers above the foliage, carrying millions of tiny flowers before it ends the tree's life by producing masses of 2cm (¾″) round fruits.

*C. umbraculifera,* Talipot Palm (southern India, Sri Lanka). A large but slow palm to 25m (80′), the trunk covered in the earlier stages with old leaf bases. The fan leaves stand upright on very long stalks and are the largest to be found on any palm, to 5m (16′) across. When the plant at last flowers, at 30 years or older, it is magnificent, with a tall branched tree-like inflorescence rising upright from the centre crown, bearing millions of cream flowers. The green fruits are 3cm (1¼″) in diameter.

*C. utan* (syn. *C. elata* ), Gebang Palm (India to Philippines, northern Australia). A large and interesting fan palm, having a heavy, scarred trunk to 20m (60′) topped with old leaf bases. The fan leaves are very large, 4–6m (12′–20′) across, with long thorny stalks. The head of creamy flowers, gradually becomes a heavy mass of fruits, olive green to black, 25mm (1″).

**Culture and habitat** Coryphas are tropical palms;

*Corypha taliera*
more resembling *C. umbraculifera.* Note the terminal inflorescence at the top of the plant.
(Photo C. Migliaccio)

*Corypha utan* (*elata*)
Gebang Palm.

*Corypha umbraculifera*
Talipot Palm. Note the terminal inflorescence at top of picture.
(Photo C. Migliaccio)

*Cryosophila argentea*
Silver Rootspine Palm.

*Cryosophila warscewiczii*
A rootspine palm from Panama.

*C. utan* occurs along the river systems of northern Australia, as well as in several other tropical regions. They are very large growers, probably best suited to parks and botanic gardens. Many palm enthusiasts are put off by the fact that they flower only once, then die. They should not be, for the flowering is a sensational event, following 20-odd years of growth. *C. utan,* although tropical, is being grown successfully in subtropical and warm temperate climates. Select a naturally moist position, or keep up the watering in drier situations. As the thorns are savage, the plant should be positioned to avoid its being a danger in traffic areas.
**Sex** Hermaphrodite. Self pollinating for seed production.
**Position indoors** No knowledge of use indoors.
**Propagation** Seeds of *C. utan* need to be fresh, and take at least 3 months to germinate. Plants are also difficult to transplant, and slow to grow.

## CRYOSOPHILA

Eight solitary species, occurring from western Mexico to northern Colombia, at low elevations in habitats ranging from dry woodlands to rainforests.
*C. argentea,* Silver Rootspine Palm (Belize). A small fan palm to 10m (30′). The trunk is covered in spines, which may turn into roots near the soil The 2m (6′) leaves are silvery beneath and have stalks 1.5m (5′) long. The inflorescences are branched and bear brown fruits 2.5cm (1″) long.
*C. warscewiczii* (Panama). This rootspine palm is very similar to the above species. The main differences are that the fan leaves are dull green above and the rootspines are longer.
**Culture and habitat** *Cryosophila* species are found in rocky limestone soils in partial shade. They are best suited to the regions near the tropics as they do not like cold, but they should also be happy in subtropical and coastal warm temperate areas if protected by other plants from cold winds and given a light tree cover. When planting, use some lime occasionally to assure an alkaline pH. Keep regularly watered until the plant is fully established and growth has doubled.
**Sex** Monoecious. Self pollinating for seed production.
**Position indoors** No indication of use indoors.
**Propagation** No information is available on germination times, although some species are being grown by some USA nurseries.

## CYPHOKENTIA

One species, *C. macrostachya,* from New Caledonia. Solitary trunk, feather leaf palm.

## CYPHOPHOENIX

Two solitary species, one in New Caledonia and the Two solitary species, one *C. elegans* in New Caledonia located in rocky rainforests, and the other *C. nucele* in the Loyalty Islands in raised coral forests.

*C. nucele* (Loyalty Islands) This species has a thick grey trunk to 6m (20′) with a crownshaft and rings. The feathery leaves are grey-green and recurved, resembling *Howea,* but sharp-pointed. The oblong 2cm (¾″) brown fruits are borne in sprays from below the crownshaft.

**Culture and habitat** *Cyphophoenix* species are fairly tough palms, coming from wooded hillsides up to 1200m (4000′) and resisting a degree of wind. While not widely grown, probably because of stock limitations, they seem quite hardy in both tropical and milder areas, with specimens growing quite well in the Royal Botanic Gardens, Sydney. They are understorey palms, so would be best suited to a home rainforest or a site under the protection of other garden palms and trees. Regular waterings and a surface mulch complete their requirements.

**Sex** Monoecious. Self pollinating for seed production.

**Position indoors** No knowledge of its use, but it could be suitable.

**Propagation** No information is available on germination times, but seeds would probably germinate best with bottom heat.

*Cyphophoenix nucele* (juvenile)
This palm grows on Lifou Island, New Caledonia.

## CYPHOSPERMA

Three solitary species, one from New Caledonia on stony hillsides to 600m (2000′) and two from Fiji to 900m (3000′) both in dense forests.

*C. balansae* (New Caledonia). A feather palm with a 6–8m (20–25′) trunk topped with a well-developed crownshaft, below which hang the inflorescence sprays, in a ring. The roundish dark fruits reach 2cm (¾″).

**Culture and habitat** *C. balansae* grows on forested slopes at an altitude of about 300m (1000′) in New Caledonia, where it is reasonably plentiful. It is quite a handsome grower, both as a potted palm and as a garden specimen. However, like so many other attractive palms from this same area, it is yet to be commercially stocked. Therefore it still remains a collector's palm, and there is a shortage of cultural information.

Although *Cyphospermas* originate from tropical islands, they are from cooler altitudes, and this allows them to tolerate more temperate climates as well as tropical ones. There are specimens to be found growing well in open positions in the Royal Botanic Gardens, Sydney. Outdoors, select a warm, shaded, well-drained position for early protection, adding some potting mix or humus to the soil for a better start. Eventually, this palm will poke through the shade to receive full sun. In nature these palms receive high and constant rainfall, so they should be kept constantly moist and not allowed to dry out. Good drainage is therefore essential, and a surface mulch is helpful to retain humidity.

**Sex** Monoecious. Self pollinating for seed production.

**Position indoors** With no commercial stocks available, no trials have been made.

*Cyphosperma balansae* (juvenile).
Another palm from New Caledonia.

*Cyrtostachys renda*
Sealing Wax Palm is very colourful.

*Cyrtostachys renda*
A large clump in Singapore.

**Propagation** No information, but bottom heat would probably assist germination. Germination times are unavailable.

## CYRTOSTACHYS

Eight solitary or clustered species, from Malaysia, Sumatra, the Borneo and New Guinea regions growing in coastal swamps *(C. renda)*, or in lowland rainforests to 500m (1500′).

*C. renda* (syn. *C. lakka)*, Sealing Wax Palm (Malaysia, Borneo). This is a beautiful, clumping feather palm to 5m (16′) high, with red crownshafts and red leaf stalks. Fruits are black on a red stalk, 10mm (⅜″).

**Culture and habitat** Sealing Wax Palms are tall, slender, clumping palms suitable only for the tropics, outdoors. In nature they grow in moist forest areas, even swamps. Their red crownshafts and leaf stalks immediately attract attention.

They are rarely seen growing outside of the tropics. Should you wish to attempt to grow one of these beautiful palms in a non-tropical-clime, select a warm, sheltered area in a suntrap, with moist soil. Specimens may be seen in conservatories in the world's botanic gardens, so a warm sunroom may be all right, otherwise a winter-heated glasshouse would be essential.

**Sex** Monoecious. Self pollinating for seed production.

**Position indoors** No information on use indoors.

**Propagation** Fresh seeds are needed, and should germinate within 2 months. Heat is beneficial. Clumps may also be divided.

## DAEMONOROPS

115 solitary or clustered species, occurring from eastern India through to southern China, then southward through to Indonesia and New Guinea. They grow as understorey plants in tropical rainforests.

*D. angustifolia* (Malaysia). This vicious climber could easily be mistaken for a species of *Calamus,* as the long climbing trunks are also very heavily covered in close, long spines. The stalks of the attractive green feather leaves are also very spiny and the midrib extending from the leaf end has spikes that hook onto anything they touch. A clump of this palm is best avoided, and in cultivation it is best confined to rainforest areas in botanic gardens. The round scaly fruits are 15mm (½″) across, with a thin layer of edible flesh.

*D. mollis* (Philippines). This rattan palm is also covered with vicious thorns. It is a creeper with feathery leaves. The long canes are used as rattan cane for furniture making.

**Culture and habitat** Closely related to *Calamus* and found in similar rainforests, these rattan palms occur in a wide range of natural habitats, each species growing under different conditions and having different light requirements, as well as soil and pH needs. Some are small weedy growers, others are

tall and sturdy tree-like climbers. The few of horticultural interest include the above two species, which follow more closely the habits of *Calamus* and require the same cultural conditions. Further information will be found under that genus. As well as open plantings, pot culture is a very satisfactory method of growing and controlling these spiky but interesting palms.
**Sex** Dioecious. Needs a male and female for seed production.
**Position indoors** No information, but they are so similar to *Calamus* that there should be no difference in indoor treatment either. I have found *Calamus* species most satisfactory and interesting under good natural lighting.
**Propagation** No information available on germination times.

## DECKENIA

One species, *D. nobilis,* from Seychelles Islands.
Solitary spiny trunk and feather leaf palm.

## DESMONCUS

61 species, from Mexico, south to Brazil and Bolivia, and Trinidad.
Clustered, spiny trunk and feather leaf palm.

## DICTYOCARYUM

Nine species, from southern Central America and north-western South America.
Solitary trunk and feather leaf palms

## DICTYOSPERMA

One solitary species, from the Mascarene Islands, possibly extinct in nature, but well represented in cultivation and native to Mauritius and Reunion Islands, where it grew throughout from sea-level to 600m (2000′).
***D. album,*** Princess Palm, Hurricane Palm (Mauritius, Réunion, Round and Mascarene Islands). This palm has a sturdy grey trunk to 10m (30′), broader at the base. This is topped with a crownshaft, which may be pale green, reddish or brown, and an attractive head of deep green feather leaves, with the newest leaf standing erect like a needle before unwinding. The flowers are reddish-yellow and the fruits purplish-black, oval and 16mm (5/8″) in length.
***D. album*** var. ***aureum*** is shorter-trunked and has orange leaf veins in juvenile plants. Not pictured.
***D. album*** var. ***furfuraceum*** is like ***D. album*** var. ***rubrum*** but with woolly white hairs. Not pictured.
***D. album*** var. ***rubrum*** has deeper green leaves, and red veins. Juvenile leaves may be red. Flowers are yellow.

*Daemonorops angustifolia*
A rattan palm from Malaysia.

*Daemonorops mollis*
A rattan palm from the Philippines.

*Dictyosperma album*
Princess Palm or Hurricane Palm.

*Dictyosperma album* var. *rubrum*
Note the red coloration.

**Culture and habitat** In appearance and climatic requirements, this palm closely resembles *Archontophoenix* and, except for the trunk, could at a glance be mistaken for that genus. The first difference you might notice would be the bulbous swelling at the base of the 10m (30′) trunk; then the dark grey colour of the trunk and its vertical splits. The crownshaft is also a different colour, being whitish-green in *D. album* (reddish in var. *rubrum). *Further, the freshly opening leaf is bound together at the edges for a longer period before fully opening. Like *Archontophoenix,* this palm is a native of rainforests, so the same growing conditions would apply. Keep out of harsh, drying winds and maintain an ample water supply. A warm position is essential.
**Sex** Monoecious. Self pollinating for seed production.
**Position indoors** Same as *Archonotophoenix.* I have observed some nice specimens growing in a high-ceilinged office, beneath skylights.
**Propagation** Propagation is from seeds, which germinate in 2–4 months. Plants require regular watering and the protection of a greenhouse.

## DRYMOPHLOEUS

15 species, from the New Guinea and Pacific area. Solitary trunk and feather leaf palms.

## DYPSIS

21 solitary or clustered species, from Madagascar and growing in lowland and mountain rainforests.
*D. pinnatifrons* (Madagascar). This slim, dainty palm was previously known as *D. gracilis.* It has a spear-like trunk, ringed and growing to 5m (15′) high. The small, shiny, deep green to reddish-green leaves resemble those of *Chamaedorea.* The solitary, usually unbranched inflorescences bear narrowly oval 13mm (½″) red fruits at maturity.
**Culture and habitat** *Dypsis* species are shade-lovers, growing under taller forest plants. They are rarely available in nurseries because stock is difficult to obtain. This also restricts the collecting of cultural knowledge. However, habitat and growing habit suggest that the main use of these palms may be as indoor pot-plants or for bush house culture. Shade, warmth and moisture seem to be essential. Their culture should be similar to that of the shade-loving *Chamaedorea* species *(C. elegans,* etc.). My photograph was taken in the bush house at Flecker Botanical Gardens, Cairns.
**Sex** Monoecious. Self pollinating for seed production.
**Position indoors** No information on indoor use recorded.
**Propagation** Seeds are said to germinate freely within 2 months, perhaps with bottom heat.

# ELAEIS

Two solitary species. *E. guineensis* occurs in humid areas of tropical Africa, where it is widespread, fringing rainforests and along watercourses in drier areas. *E. oleifera* occurs in poorly drained and swampy equatorial areas of Central America and northern South America.

*E. guineensis,* African Oil Palm (tropical western and Central Africa). This is a handsome feather palm eventually, with a rough trunk to 10m (30′) tall. The deep green leaves are about 4–5m (12′–16′) long. The 4cm (1½″) fruits, borne in large clusters, are red to black.

*E. oleifera,* American Oil Palm (equatorial Central and South America). This species has a heavy trunk that may first lie along the ground and then rises to 2m (6′) high, with clinging leaf bases. The deep green feather leaves are full and broad. The 25mm (1″) dark orange fruits are in congested heads.

**Culture and habitat** These are the valuable oil palms, fast becoming a crop almost as valuable as the coconut. *E. guineensis* is being planted in many tropical areas of the world for its palm oil, used in the manufacture of food, and also cleansing and lubrication products. This palm is also a handsome specimen plant for single and avenue planting. *E. oleifera* (syn. *Corozo oleifera)* is likewise useful and ornamental in the warmer tropical South American regions.

Besides being grown as an oil crop, *Elaeis* is also used locally for wine, thatch and building materials.

*Elaeis* species are quite popular as garden specimens and potted palms. They are best in tropical and subtropical regions, where *E. guineensis* appears to be the hardier or more often grown species. It can also be grown in the warmer zones, away from frost or excessive cold. Being large growers, it is hard to provide *Elaeis* species with a great deal of protection, but in the cooler regions it may be best to plant them in the cover of other palms and tall shrubbery, to begin, although they will eventually grow up out of these. They require regular water. In warmer areas, they will tolerate salty coastal conditions.

**Sex** Monoecious. Self pollinating for seed production.

**Position indoors** Not used indoors, to my knowledge, but they are popular outdoor pot-plants.

**Propagation** The seeds have a very hard shell. Cracking or soaking in hot water for 2 weeks and bottom heat may be beneficial. Germination may take 2–4 months, and a moisture-retaining medium such as composted sawdust is useful.

*Dypsis pinnatifrons*
A small shade-loving palm (centre), showing the long, slender trunk.

*Elaeis guineensis*
African Oil Palm.

# ELEIODOXA

One to five species, from southern Thailand and Malay Peninsula, to Sumatra and Borneo.

Clustered spiny trunks and feather leaf palms.

*Elaeis oleifera*
American Oil Palm.

*Euterpe edulis*
Assai Palm, the foliage is eaten as food.

## EREMOSPATHA

12 or fewer species from Central West Africa.
Clustered, climbing rattan trunks and feather leaf palm.

## EUGEISSONA

Six species, from the Malay Peninsula and Borneo.
Clustering spiny trunks and feathered leaves.

## EUTERPE

28 solitary or clustered species, occurring from the West Indies and Central America southward through Brazil to Peru. They grow in habitats ranging from lowland rainforests to mountain forests, up to 3000m (10000′). They may also be found growing in swamps. They either have solitary trunks or grow in clusters. The round fruits may be purple, white or black and from 10–25mm (½″–1″) long.

*E. edulis,* Assai Palm, Palmito (Brazil). This slender palm has a narrow trunk to 30m (100′) high, with a prominent crownshaft and a head of fine arching leaves. The fruit is small and round. There is a drink, quite popular in Brazil, called 'assai' that is prepared from the fruit. The pulp from the trunk is also eaten in Parana. The leaf crown is harvested under the name of "palmito".

*E. purpurea* (Colombia). In this species the leaf stalks and crownshaft have a blackish-purple colouring, particularly in the juvenile plant, making it very attractive.

*E.* sp. (tropical America). This is a handsome cluster palm with dainty feather leaves, nicely displayed above a deep green crownshaft.

**Culture and habitat** Euterpes are native to Brazil and the Caribbean where they have been used as food for many years. They occur in areas of high rainfall, in rainforests, and some grow on steep mountain slopes. They are fast-growing and slender.

*Euterpes* are not popularly grown out of their habitat or botanic gardens, maybe because of their height and slenderness. They would require a warm, frost-free climate, away from cold winds and salt burn. They need some foliage protection from hot burning sun, as their foliage is quite lush and tender. In nature they get this protection by growing in elevated rainforests. A surface mulch and regular water will help to keep them looking healthy. It would be beneficial also to spray their foliage regularly.

**Sex** Monoecious. Self pollinating for seed production.

**Position indoors** As *Euterpes* have foliage with a nice, soft green appearance, they are quite attractive as a potted palm. Select a bright and warm position and keep amply watered but do not allow the soil to dry out. Take the plant outdoors onto a warm veranda if it becomes distressed.

At one stage, a large number of *E. edulis* plants were grown in Australia and sold for many years in temperate climates, where chain stores marketed

them as indoor plants. It is a tragedy that none of these seem to have survived after being transplanted out of doors in those cooler regions. There are survivors still in the subtropics, though.
**Propagation** Seeds germinate in about 1 month and growth is rapid.

## GASTROCOCOS

One solitary species, growing in Cuba, in limy clay soils.
*G. crispa,* Cuban Belly Palm (Cuba). The unusual shape of the 9–14m (30′–45′) trunk of this palm immediately catches the eye, swelling as it rises and variably covered in bands or rings of spines. This is topped with a dense crown of feathery fronds, with narrow leaflets that are glossy green above and blue-green beneath. The inflorescence hangs down and when the fruits develop they resemble a hanging bunch of grapes, 24mm (1″) long and yellow-brown.
**Culture and habitat** *Gastrococos* species are very similar to *Acrocomia,* and in fact were once confused with them, and require the same cultural conditions. Their growth can be slow, sometimes in spurts and rests.
**Sex** Monoecious. Self pollinating for seed production.
**Position indoors** No information on use indoors.
**Propagation** Germination takes 4–6 months and is difficult. Seeds are best either filed to weaken the shell, or soaked in water and kept warm for a couple of weeks.

## GAUSSIA

Four solitary species, which now include *Opsiandra,* growing in Guatemala, Cuba and Puerto Rico, southern Mexico, Belize, on limestone hillsides, occasionally limestone lowlands and forest floors of highlands.
*G. attenuata,* Llume Palm (Puerto Rico). This palm grows 30m (100′) on a tapering slender trunk about 15cm (6″) wide at the ground and narrowing as it rises to the crown of sparse feather leaves. Around the base there is a prominent network of surface roots. The small branching inflorescence near the crown produces a good cluster of 1cm (⅜″) round orange fruits.
*G. maya* (syn. *Opsiandra maya),* Maya Palm (Belize, Guatemala). This species has a trunk to 20m (60′) which may be covered in inflorescences; it is topped with almost plume-like feather leaves which have numerous leaflets displayed in different planes. Flowers are greenish and fruits red.
*G. princeps,* Sierra Palm (Cuba). A palm with a tapering trunk to 15m (50′) high, and exposed roots at the base, its feathery leaves, 2m (6′) long, are arching and dainty. The hanging inflorescence stem is chain-like, and fruits are 1cm (⅜″) long, purple to red.
**Culture and habitat** *Gaussias* inhabit limestone cliff areas and steep slopes. Being tall and slender, they

*Euterpe purpurea* (juvenile)
This species has purple coloration.

*Euterpe* sp.
A clustering species from tropical America.

*Gastrococos crispa*
Cuban Belly Palm.

appear to hang onto the ground surface by their network of surface roots. Even when they are planted in richer soils, these surface roots still appear. The small leaf crown is disappointing, giving the palm a starved appearance.

Although they come from tropical areas *Gaussias* appear to be quite tolerant of temperate climatic conditions and should survive along the coastal belt of the warm temperate zone or subtropics if protected from cold winds and frosts. The warmer humid areas are of course ideally suited to these palms. Originating from limestone soils, they would appreciate an occasional liming if grown in acid or clay soils. They tolerate poorer soils, but grow into nicer specimens if looked after.

**Sex** Monoecious. Self pollinating for seed production.

**Position indoors** No knowledge of indoor use.

**Propagation** Seeds germinate in about 2 months. Plants are cold-tender.

## GEONOMA

75 solitary or clustered species, occurring from Mexico to Brazil and Bolivia, as forest floor plants. They populate low to high elevations.

*G. apetiolata* (tropical America). Little is known about the growth habit, inflorescences or fruiting of this attractive little palm growing at Fairchild Tropical Gardens, Florida.

*G. interrupta* (Caribbean). This species has slender,

*Gaussia attenuata*
Llume Palm.

*Gaussia maya*
Maya Palm.

multiple trunks from 1-7m (3′-24′) high, shiny brown and ringed. The feather leaves are unevenly divided and glossy green. The 60cm (24″) inflorescences, bear round pointed black fruits 6mm (¼″) long. The leaves may be eaten by the locals as 'cabbage'.

**Culture and habitat** *Geonoma* species are low, slender palms of the jungle understorey, with ringed trunks and no crownshaft. The small head of leaves may be feathery, or have any number of broad segments on each side of the midrib. Many species grow at elevations, suggesting some tolerance of cooler conditions.

Being understorey palms, *Geonomas* would be best planted in tropical or subtropical group plantings, in home rainforests or otherwise in shaded corners. Those species that come from cooler elevations may also be suitable for the warm temperate zone. Here it would be best to select a warm shaded area protected from cold winds a shaded suntrap. A surface mulch and ample regular watering will provide humidity. They make good potted palms in the right positions.

**Sex** Monoecious. Self pollinating for seed production.

**Position indoors** No information on indoor use is available, but worth trying as they are suited to pot culture.

**Propagation** Seeds should germinate in 1–2 months.

## GONIOCLADUS

One species *G.petiolatus,* from Fiji.
Solitary trunk and feathered leaves.

## GRONOPHYLLUM

33 solitary or clustered species, which now include *Nengella* and *Kentia,* occurring from Sulawesi through to New Guinea and northern Australia, in habitats varying from sea-level forests to mossy mountain forests; in habit they range from understorey plants to emergent trees.

*G.* sp. (previously *Nengella* sp., New Guinea) from Cairns Botanic Gardens. This New Guinea species has multiple slender trunks to about 5m (16′) high. The segmented green leaves with brownish markings have cut off ends. The fruiting spikes are small and bear somewhat oval fruits about 14mm (½″) long which may be bright red or purplish-black.

**Culture and habitat** These include small palms from the New Guinea undergrowth in many ways resemble *Chamaedoreas.* They require warm situations with shade. Not much is known about them and because supplies are limited there is little information about their culture in temperate climates away from their tropical habitats. They would be best planted amongst other palms and plants that can shade them. The home rainforest, a shady corner, or a fernery or bush house would be ideal. A surface mulch and regular waterings would supply the humidity they need. They are also suitable for pot culture.

*Gaussia princeps*
Sierra Palm.

*Geonoma apetiolata*
This species shows signs that it will be interesting.

*Geonoma interrupta*
The young leaves are eaten as food.

*Gronophyllum* sp.
At left, pictured with *Chamaedorea brachypoda* on the right.

**Sex** Monoecious. Self pollinating for seed production.
**Position indoors** No information available, but would seem suited to indoor culture.
**Propagation** Seeds germinate best with bottom heat, and take 4–5 months. The plants are slow-growing and would be best kept in glasshouses or greenhouses in warmer climates.

## GUIHAIA

Two species, from southern China and North Vietnam.
Clustering trunks and fan leaves.

## GULUBIA

Nine solitary species, extending from the Moluccas to New Guinea, northern Australia and the Pacific Islands as far as Fiji, and growing in lowland to mountain rainforests.
*G. costata* (Cape York Peninsula, New Guinea). This feather palm has a tall ringed grey trunk to 20m (60′) or more, with a pale green crownshaft. This is topped with a handsome head of narrow leaves much resembling a Bangalow Palm, but with drooping leaflets. The inflorescence bears cream flowers and then oval maroon fruits 15mm (½″) long.
**Culture and habitat** *G. costata* is a handsome palm, not very well known as yet, and suiting the tropics best. In nature it grows in stands in rainforests, being tall enough to peep above the trees.
*G. cylindrocarpa,* growing at Fairchild Tropical Gardens, Florida. This juvenile plant is impressive, but little is known of its matured growth at this stage.
It deserves more attention than it receives at present. It is a shame that it is restricted to the tropics; the handsome specimens I photographed were thriving in a Cairns garden. I have not seen any attempt to grow this native palm in any of the botanic gardens.
Being sensitive to cold, it is only recommended to be planted in tropical gardens, in warm situations that provide shade when it is young. The plant will soon grow upwards to take full sun. If you wish to try a *Gulubia* in a more temperate climate, select a warm suntrap area with initial shade and continuing protection from cold winds.
*Gulubias* require a lot of water and will even grow in swampy positions. Add some humus or potting mix to the soil before planting. Do not allow to dry out.
**Sex** Monoecious. Self pollinating for seed production.
**Position indoors** There is no information available on indoor use.
**Propagation** Seeds germinate in about 3 months. Plants should be transplanted when quite young to avoid root damage. Bottom heat for the seeds and glasshouse protection for the young plants would be an advantage.

## HALMOOREA

One species, *H. trispatha,* from Madagascar.
Solitary trunk and feathered leaves.

## HEDYSCEPE

One solitary species from Lord Howe Island, New South Wales, growing on high cliffs over the sea and on exposed ridges at 600–750m (2000′–2500′).
*H. canterburyana,* Umbrella Palm (Lord Howe Island). The grey, close-ringed trunk to about 5–10m (15′–30′) high is topped with a short, swollen silvery green crownshaft. The stiff-looking, deep green feather leaves resemble those of *Howea belmoreana,* and also recurve. There is a bunched skirt of flowers from below the crownshaft, followed by egg-shaped 4cm (1½″) oval deep red fruits.
**Culture and habitat** *H. canterburyana* at one time used to be included in the genus *Kentia* but because of seed differences was given its own genus. It differs only little and mainly by altitude from its Lord Howe Island neighbours *Howeas,* and its growing conditions are similar. See *Howea* for cultural information.

It grows in forests and on windy hillsides at an altitude of about 600–750m (2000′–2500′) the trunk

*Gulubia costata*
A handsome palm for the tropics.

*Gulubia cylindrocarpa*
This juvenile palm is growing at Fairchild Gardens, Florida, USA.

*Hedyscepe canterburyana*
Umbrella Palm is native to Lord Howe Island.

*Heterospathe elata*
Sagisi Palm.

*Heterospathe minor*
A dark-stemmed species from the Solomon Islands (foreground).

height varying according to the surrounding vegetation. It also grows on cliffs overlooking the sea, where it broadens up to resist the strong salt winds.

This palm prefers temperate climates, not far removed from the coast; it is difficult to grow in the tropics.

**Sex** Monoecious. Self pollinating for seed production.

**Position indoors** Said to be as good as *Howea (Kentia)* indoors, tolerating positions with good to poor natural light. I have always found that the better the light, the longer it lasts indoors.

**Propagation** Seeds are slow to ripen, taking about 4 years. Germination is also slow and erratic, taking up to 12 months.

## HETEROSPATHE

32 usually solitary species, occurring from Indonesia to the Philippines and New Guinea to the Solomons. They are mostly understorey palms of lowland to mountain rainforests. Fruits are about 12mm (½″) long, oval and orange to red.

*H. elata,* Sagisi Palm (Philippines, Ambon). This palm grows to 15m (50′), with a smooth ringed trunk which is broader at the base. The feathery green leaves are 3m (10′) long and arch gracefully. New leaflets are pinkish brown. The inflorescence is 1m (3′) long and the white to red fruits are 12mm (½″) long and oval. Poor resistance to cold temperatures and cold winds.

*Heterospathe negrosensis*
A slender palm for shade.

*H. minor* (Solomon Islands). This species has a slender trunk and graceful deep green feather leaves with a dark stalk.
*H. negrosensis* (Philippines). An interesting species with a slim trunk and handsome feather leaves; it differs from *H. elata* in having blackish green leaf stalks.
*H. sibuyanensis* (Philippines). A taller grower to 10m (30′) preferring shade. The green feather leaves have small brown scales on the undersides. The inflorescence is branched and later bears pointed conical fruits 12mm (½″) long.
**Culture and habitat** *Heterosphathe* species come from the Philippines, New Guinea and the Pacific Islands. They are tropical and sensitive to cold. In their natural habitat they are understorey plants, with a few of the taller species breaking through to receive direct sunlight. They grow best in tropical and subtropical areas, but there are specimens growing in botanic gardens in warm temperate areas, as well as potted glasshouse specimens. Select a warm position, preferably protected from cold winds, or otherwise a rainforest type situation or a suntrap. Build up the soil by adding compost or potting mix to ensure good drainage. Water regularly and feed occasionally. Apply a surface mulch for added moisture retention and humidity.
**Sex** Monoecious. Self pollinating for seed production.
**Position indoors** No information.
**Propagation** The seeds germinate within 2 months, and prefer bottom heat. Keep young plants in a warm position.

*Heterospathe sibuyanensis*
A taller Philippine species.

## HOWEA

Two solitary species from Lord Howe Island. *H. forsteriana* grows in sandy lowland forests, while *H. belmoreana* can be found scattered through the same forests, but is more abundant at higher elevations, to 450m (1500′).
*H. belmoreana,* Sentry Palm (Lord Howe Island). The slender ringed trunk, wider at the base, grows to about 6m (20′). The deep green recurving leaves have their leaflets held upright in a V formation. Instead of a crownshaft there is a mat of fibres and from this hang the 2m (6′) long, cat-tail-like inflorescences. The 3cm (1″–1¼″) fruits are shiny green, later becoming brownish.
*H. forsteriana,* Kentia Palm, Thatch Palm (Lord Howe Island). The ringed trunk is slender and grows to 15m (50′) high, bearing a head of deep green leaves which droop at the tips as they age. This species too has a mass of brown fibres instead of a crownshaft, from which hang the cat-tail inflorescences. Fruits are oblong, 3–4cm (1¼″–1½″) long, reddish brown.
**Culture and habitat** Both species are found (at different altitudes) on Lord Howe Island, along with *Hedyscepe canterburyana.* Of the two, *H. fosteriana* is the more popular, and is widely sought after in most countries of the world. This makes the island's seed crops a valuable export commodity, as every available seed is harvested for the world market. Their

*Howea forsteriana*
A leading indoor potted palm.

*Howea belmoreana*
The leaves of this species curve downwards.

*Howea belmoreana*
Sentry Palm of Lord Howe Island.

popularity comes from their hardiness as an indoor palm and for this purpose they are the number one commercially used palm. Seeds take years to ripen.

This valuable palm may be successfully grown in frost-free gardens in the coastal districts of the temperate and subtropical zones. They do not like dry, hot winds nor too much cold. Neither will they tolerate the humid tropics as successfully as the cooler zones. It is possible to grow them in the tropics, but they suffer from more leaf markings.

It is best to start them shaded from the sun to keep them from stunting. They soon grow up to become sun-tolerant. In warmer climates, such as the tropics or subtropics, they may prefer some shading from the hot sun or drying winds. This may be done with trees or by planting them on the eastern side of a building. Before planting, dig in some compost over 1–2m (3′–6′). Water regularly, but select a well-drained position.

**Sex** Monoecious. Self pollinating for seed production.

**Position indoors** The popularity of *Howeas* derives from their hardiness indoors. However, the better the light, the better growth will be. Feed occasionally with a complete plant food, particularly when the plant is making new growth. Hose the dust off on dull days, or place the plant outside on rainy days, but do not allow the sun to damage the leaves by leaving it out too long.

**Propagation** Always use properly ripened and fresh seeds, which take from 2 months to germinate. Germination is more even if bottom heat is used. Seedlings are very slow, taking many years to reach a commercial selling size.

## HYDRIASTELE

Eight solitary or clustered species from New Guinea and northern Australia, occurring in lowland to upland tropical rainforests.

*H. microcarpa* (Lae, New Guinea). This juvenile specimen is growing in the dome at Brisbane Botanical Gardens at Mt Coot-tha. A tall grower with feather leaves and leaflets with cut-off ends. Appears to resemble *H. microspadix.*

*H. wendlandiana* (northern Australia). This tall, slender palm has clumps of ringed trunks 7–20m (20′–60′) tall. These are topped with a whitish crownshaft and a sparse head of very recognisable feather leaves. The spikes of cream flowers are followed by small red fruits 9–10mm (3⁄8″) long.

**Culture and habitat** *Hydriastele wendlandiana* is restricted to coastal areas of the Northern Territory and Cape York Peninsula, where it may be found in tropical rainforests or rich swampy areas. It is sensitive to cold and will die if subjected to frosts. Specimens are growing in many of the warmer botanic gardens including the subtropics. Other species of *Hydriastele* are native to New Guinea, which is more tropical. Fruits are reddish to purplish.

This warmth-loving palm is best grown in the tropics. If you wish to attempt growing it in a cooler

district, select a consistently warm, protected position, a suntrap area with overhead shade, or a warm rainforest situation. Dig in some compost or potting mix to richen the soil, and keep up regular waterings. While this palm requires semishade in its early life, it will eventually tolerate full sun, but not cold winds. It is also useful as a potted palm for protected corners, or for indoor use.
**Sex** Monoecious. Self pollinating for seed production.
**Position indoors** Select a warm position out of draughts, with good natural light. No information is available about its hardiness indoors.
**Propagation** Propagation is from seeds or by division of suckers. Seeds need to be fresh. They require bottom heat and take 3–6 months to germinate.

*Howea forsteriana*
Kentia Palm or Thatch Palm.

## HYOPHORBE

Five solitary species from the Mascarene Islands, where they used to grow both in the valleys and in mountain forests to 700m (2300′). Now they are almost extinct in nature, but are cultivated everywhere.
*H. indica* (Mascarene Islands). A taller-growing species to (15m [50′]). The trunk is twisted but does not bulge as in other species. It is topped with a bulging crownshaft and six to eight stiff feather leaves about 2m (6′) long. Fruits are round and scaly. This species is now almost extinct in the wild with only cultivated specimens remaining.

*Hydriastele wendlandiana*
A showy clumping palm.

*Hydriastele microcarpa* (juvenile)
An introduction from New Guinea.

*Hyophorbe indica*
A tall-growing species.

*Hyophorbe lagenicaulis*
Bottle Palm, observe the bulbous trunk.

*H. lagenicaulis,* Bottle Palm (Mascarene Islands). The interest in this dainty palm is the bottle-shaped, grey trunk, which is widest at the base, then tapers upwards to the attractive crownshaft, which forms the neck of the 'bottle'. There is a sparse head of stiff, smooth, recurving green leaves with very short stalks. The flowers are borne on horn-like inflorescences below the crownshaft, and the fruits are orange to black, 3cm (1¼") long and oblong.

*H. verschaffeltii,* Spindle Palm (Mascarene Islands). This stiffly attractive palm has recurving leaves with short stalks. The trunk, which is golden at the base, is triangular on young plants; later it bulges out up to the crownshaft, which then narrows again to the leaf bases. The flowers are borne from what look like rhinoceros horns below the crownshaft. The fruits are red to black, 18–20mm (¾") long and oblong.

**Culture and habitat** In nurseries these palms will more often be seen under the name of *Mascarena.* They originate from the Mascarene Islands, and they have a tolerance to salt winds.

Their attraction lies in the shapes of the trunks, which are most attractive and unusual, and always eye-catching. They will stand full sun, but require ample moisture and care. They give best results, and look most attractive, if well cared for and regularly nurtured, and can become very straggly if totally neglected in dry positions. Dig some indoor potting mix or compost over 1–2m (3′–6′) into the soil before planting. A warm to sheltered position is preferable. Water regularly.

**Sex** Monoecious. Self pollinating for seed production.

**Position indoors** These palms are not usually used indoors, but they are most attractive where there is room for them to spread. A sunroom would be best, or a sheltered veranda.

**Propagation** Propagation is from seeds, which germinate within 3 months. In cooler climates keep plants in warm sites, or in a glasshouse.

## HYOSPATHE

About 17 species, from Costa Rica to Peru.

Solitary or clustered trunk palms with feathered leaves.

## HYPHAENE

Ten solitary or clustered species, distributed from drier Central Africa through a coastal fringe to southern India, also western Madagascar. They grow in arid regions that have ground or surface water.

*H. petersiana* (syn. *H. ventricosa),* Elala Palm (southern Africa). The rough trunk is covered in old leaf bases and grows to a height of 10m (30′) if it remains single. The fan leaves, resembling those of *Sabal,* are blue-green with long purple spiny stalks. The yellow flowers are followed by a heavy bunch of yellow pear-shaped fruits 5–6cm (2″–2½″) long.

*H. ternifolia* (Africa). This species has a suckering trunk, and the handsome blue-green fan leaves have blackish bases.

**Culture and habitat** *Hyphaene* species are relatively untried in commercial nurseries. They come from Africa, Madagascar and India, where they mostly occur in hot, dry, poorer areas. They are generally referred to as 'doum palms'. Most tropical botanic gardens around the world have a few specimens. They are usually a botanical curiosity.

Hyphaenes should grow successfully in most warm and even dry areas which are not subject to heavy frosts. They require an open, sunny position, and need attention until established. Water during drought periods.

**Sex** Dioecious. Need a male and female for seed production.

**Position indoors** No information.

**Propagation** Seeds take up to 6 months to germinate, but may be hastened a little by cracking, or even removing, the hard shells. The taproot goes 40–50cm (16"–20") down into the soil, so very deep containers are necessary for raising them in pots.

## IGUANURA

About 18 species, from southern Thailand, southward to Borneo and Sumatra. Solitary or clustered trunks and undivided (sometimes feathered) leaves.

*Hyophorbe H. verschaffeltii* (right rear) *H. lagenicaulis* (left) with fruit.

*Hyphaene petersiana*
The colourful Elala Palm.

*Hyophorbe verschaffeltii*
Spindle Palm. A feature planting in Singapore.

*Hyphaene ternifolia*
A Doum Palm from Africa.

*Iriartea gigantea* (juvenile)
A stilt palm from Costa Rica.

## IRIARTEA

Seven solitary species described, but probably only two, from Costa Rica to northern South America, in rainforests from the lowlands to 1200m (4800′).
*I. gigantea,* Stilt Palm (Costa Rica). This palm is a tall grower 10–25m (30′–80′). On a mature plant the stilt roots may be 1m (3′) high. The mature leaves, about six per plant, have multiple narrow leaflets in a plumose formation. The round fruits are 25mm (1″) in diameter and are yellow at maturity. This species grows on hillsides.
**Culture and habitat** *Iriarteas, Socrateas* and other stilt palms usually inhabit wet tropical or equatorial areas with high rainfall. They are suited to large reserves and suitable parks, in tropical areas but are hardly ever grown in private gardens. Younger plants may be grown in pots. Refer to *Socratea* for further information. The timber of *Iriartea* is very hard and rot-resistant.
**Sex** Monoecious. Self pollinating for seed production.
**Position indoors** No information available.
**Propagation** No precise information available, but bottom heat and soaking may be helpful.

## ITIARTELLA

Two species, from southern Central America and north-western regions of South America.
Usually clustered trunks and feathered (broad leaflets) leaves.

## ITAYA

Two species, from Amazonian areas of Peru and Brazil.
Solitary trunks and fan leaves.

## JESSENIA

One species, *J. bataua,* with two subspecies, from Panama, Trinidad and northern areas of South America.
Solitary trunks and feathered leaves.

## JOHANNESTEIJSMANNIA

Four solitary species, from Malaysia, Sumatra and western Borneo, growing as understorey plants in tropical rainforests, but avoiding overly wet soils.
*J. altifrons* (Sumatra, Malaysia). An eye-catcher because of its large undivided leaves. These catch litter and direct it down around the trunk, which may remain below ground level or rise to 2m (6′). The large 2m (6′) paddle-shaped leaves are used as

thatching. This palm has a short 15cm (6") flower stalk and round warty brown 3.5cm (1½") fruits.
**Culture and habitat** Outside their natural habitat in the shade of equatorial jungles, these palms are likely to be seen only in tropical botanic gardens, or as glasshouse specimens in enthusiasts' collections. These very eye-catching palms were formerly called *Teysmannia.*

*Johannesteijsmannias* would be best suited to tropical shaded gardens, or home rainforests, with ample moisture. They could also be grown as tub specimens under suitable conditions, or in glasshouses beyond the humid tropical regions.
**Sex** Hermaphrodite. Self pollinating for seed production.
**Position indoors** No information available because of lack of stock for trial.
**Propagation** No information available on germination times.

# JUANIA

One species *J. australis,* from the Juan Fernandez Islands. Solitary trunk and feathered leaves.

# JUBAEA

One solitary species, from warm temperate Chile, growing in dry ravines and on ridges amongst woodland scrub.
*J. chilensis,* Coquito Palm, Chilean Wine Palm (Chile). The massive grey trunk, to 25m (80') tall and 1.3m (4') broad, is topped with a massed head of stiff feather leaves, dull green above and greyish beneath. The thick, heavy inflorescence hangs down. The orange-yellow fruits are roundish with a point, 3–3.5 cm (1¼") in diameter.
**Culture and habitat** *Jubaea* is suited to zones around the warm temperate zone. It can be a little touchier to grow in the tropics, while in cooler regions it will tolerate some frost. It is closely related to *Butia and Syagrus,* which are also extremely hardy. Another close relative is *Jubaeopsis,* which has multiple trunks. In its native Chile, *Jubaea* used to be felled for its sap, which yields palm honey and alcohol by distillation. Its usefulness to humanity, led to its existence being threatened, but it is now protected. This palm may be grown in cooler regions, but specimens in those areas are more limited. A large grower, it is hardy and will adapt to cold and dry conditions. It is tall and will stand above most other vegetation. Maintain watering until the plant is fully established and healthy.
**Sex** Monoecious. Self pollinating for seed production.
**Position indoors** No information is available.
**Propagation** Germination is slow, taking 6–16 months.

*Johannesteijsmannia altifrons*
(juvenile)
The leaf is unsplit.

*Jubaea chilensis*
The massive head of leaves stands above other palms.

*Jubaeopsis caffra*
Pondoland Palm or Kaffir Palm

*Kentiopsis oliviformis*
A beautiful but endangered ornamental palm.

## JUBAEOPSIS

One clustered species, from South Africa, growing on the steep northern rocky banks of two rivers near the coast. Closely related to *Butia* and *Jubaea.*

*J. caffra,* Pondoland Palm, Kaffir Palm (South Africa). This is a most unusual and little-known palm, but specimens are growing in many botanic gardens. Very similar to the Coconut Palm, it generally has clustered trunks, but has been known to reach a height of 6m (20′). The 4m (12′) feather leaves are tightly packed with narrow leaflets which are uneven in length. The hanging inflorescence has a long stem and the fruits resemble coconuts but are only about 3cm (1¼″) long.

**Culture and habitat** *Jubaeopsis* was originally considered to be a species of *Jubaea.* In South Africa it follows the course of rivers, growing on the shady side of elevated cliffs. While not growing on the river bank, it still prefers regular water, but is also hardy enough to withstand drought.

There is only scant information available on preferred growing conditions, but it may be assumed from its natural range that this palm is suited to the tropics and subtropics. However, there is a report that a cold spell in California killed several *Jubaeopsis* plants. A specimen growing in Brisbane was doing well in a hot exposed position that seemed drier than its reported natural habitat. Another in Hawaii as also growing well and receiving more water than the Brisbane specimen. With so few *Jubaeopsis* specimens in cultivation to observe, its cultural requirements can only be guessed at.

In the natural habitat it appears that the plants are exposed to sunlight, and have the benefit of good drainage combined with a plentiful water supply in an area of regular rainfall. It would be best to match these conditions, and to attempt growing this palm only in a frost-free area.

**Sex** Monoecious. Self pollinating for seed production.

**Position indoors** No information available.

**Propagation** Reports indicate that seed is difficult to germinate, but once established, seedlings are sturdy and hardy.

## KENTIOPSIS

One solitary species from New Caledonia, where it grows along the banks of streams.

*K. oliviformis* (New Caledonia). This species has a closely ringed trunk 10m (30′) or taller, topped with a long, smooth light green crownshaft. The feathery leaves have narrow leaflets which tend to droop downwards. The inflorescence, growing from below the crownshaft, is broom-like and held upwards at about 45°. It later bears olive-shaped fruits 2cm (¾″) long.

**Culture and habitat** *Kentiopsis* is now very rare, and few plants are left in the wild, because it competes for the flat land now needed for other purposes in its native New Caledonia. It also grows along the edges of creeks and streams, and seems to require a constant and regular water supply.

Being tropical, it is easily grown in the warmer zones, provided it receives a constant supply of water. It also seems to grow quite well in protected positions in warm temperate coastal areas, and there is a specimen in the Royal Botanic Gardens, Sydney.

Because *Kentiopsis* is rare, it is most unlikely that plants will be offered for sale in nurseries. Should you be fortunate enough to obtain one, plant it in a warm sheltered position, remembering that it will grow over 10m (30′) in height, and give it protection from cold and strong winds. Add some humus or potting compost to the soil, and water regularly.

**Sex** Monoecious. Self pollinating for seed production.

**Position indoors** Little is known of indoor use.

**Propagation** Reports indicate that good viable seeds produced patchy results. No germination times are available.

## KERRIODOXA

One species *K. elegans* from Thailand. Solitary trunk and fan leaves.

## KORTHALSIA

26 clustered species, occurring from India to China through to Burma, Indonesia and New Guinea, in lowland and hill tropical rainforests, and both primary forest and regenerated areas.

*K. laciniosa* (syn. *K. teysmanni),* Ant Rattan (South-East Asia). A climbing rattan palm armed with hook-spines on the 15m (45′–50′) cane-like stems. The unusually shaped leaflets have a thorny stalk, and a thorny whip-like extension which attaches to supports to help the plant climb. The 1.5m (5′) branched inflorescences later bear golden scaly oval fruits 2cm (¾″) long.

**Culture and habitat** A tropical jungle palm, much resembling *Calamus,* but not used to make cane furniture. In horticulture it is probably only a collector's palm for parks, gardens and pot culture, as is *Calamus.*

If wishing to grow a *Korthalsia,* a shady position near a tree in a warm frost-free climate is essential, and ample water must be given when needed. The specimen photographed was growing in Brisbane Botanic Gardens. As a potted palm in the temperate zones it may need a warm bush house or glasshouse. See *Calamus* for further information.

**Sex** Hermaphrodite. Self pollinating for seed production.

**Position indoors** No information available, but worth a trial, as *Calamus* is excellent indoors.

**Propagation** Seeds germinate better if fresh, but no times are available.

*Pinanga javana*
Nice variety from West Java.

*Laccospadix australasica*
Atherton Tableland Palm. A mature specimen.

*Laccospadix australasica*
A red-leaved species, common in the species.

*Laccospadix australasica*
This plant has a single trunk. There is variation within the species.

## LACCOSPADIX

One clustered species, from north-eastern Australia, growing in rainforests on tropical tablelands.

*L. australasica,* Atherton Tableland Palm, Queensland Kentia (Australia). Often purchased as a single-stemmed plant less than 1m (3′) high, this species may soon develop new trunks from the base, each of which will grow to about 3m (10′) high but only 5cm (2″) wide. The trunks are crowned with deep green leaves. Spent leaf stems are best removed. The oval 14mm (½″) yellow fruits change to red when mature.

**Culture and habitat** *Laccospadix* originates from the Atherton Tableland area where it grows on the mountain slopes and ridges in rainforests as an understorey palm. Little-known up to a few years ago, it is now one of most popular palms for indoor use in Australia. It closely resembles a *Howea* in its juvenile form, and could easily be mistaken for one. In its later growth it may become rather different from *Howea* developing multiple trunks which are much slimmer and shorter than in *Howea,* or it may remain as a solitary-trunked palm. Perhaps future research into this species will separate it into three or four distinct varieties or species. Some of the variants are pictured.

Planted in the garden it required total shade, and a tropical or temperate climate with cool, but not frosty nights. There are specimens growing outdoors in the major botanic gardens. The foliage should be regularly hosed and the soil kept moist. Dig in some indoor potting compost spread over 1–2m (3′–6′) before planting, and supply a surface mulch to maintain humidity. Keep moist.

**Sex** Monoecious. Self pollinating for seed production.

**Position indoors** *Laccospadix* stands up quite well indoors if kept regularly moist, but it should not stand in a saucer of water. The dark foliage contrasts well against light walls. An occasional leaf will die and may be removed. Side shoots will appear to fill the pot. Multiple plantings are also effective and are used a lot by the nursery trade. When first put indoors it will quickly increase growth by about 25 per cent but it then settles down.

**Propagation** Propagation is from seeds, which take 2–12 months to germinate (if bottom heat is used, 2–3 months). Young growth is slow.

## LACCOSPERMA

Seven species, from West Africa.

Clustered climbing rattan palms, with feathered leaves.

## LATANIA

Three solitary species, from the Mascarene Islands growing on cliffs and in ravines near the sea. They are now almost extinct there, but widely cultivated elsewhere.

*L. loddigesii,* Blue Latan Palm (Mascarene Islands). This species has a 15m (50′) trunk topped with very stiff blue-grey fan leaves which are white-scurfy beneath. Fruits are purplish black, 5–6cm (2″–2½″) long.

*L. lontaroides,* Red Latan Palm (Mascarene Islands). The grey trunk, swollen at the base, grows to 16m (60′) and is topped with stiff fan leaves, grey-green with red veins and with white scurf at the base. Juvenile growth is reddish. Fruits are greenish-brown, 5–6cm (2″–2½″) long.

*L. verschaffeltii,* Yellow Latan Palm (Mascarene Islands). The rough trunk to 15m (50′) is topped with stiff fan leaves, light green with a white scurf beneath, with yellow stalks. Fruits are greenish-brown, 5–6 cm (2″–2½″) long.

*L. loddigesii* x *L. verschaffeltii.* This sturdy cross has reddish brown leaf stalks.

*L.* sp. This attractive form has pink leaf stalks and bluish fan leaves.

**Culture and habitat** *Latanias* are handsome fan palms from the Mascarene Islands, where they grow overlooking the sea in exposed conditions. They appear quite hardy and able to withstand dry spells. Although from tropical areas, they will grow quite well in warmer temperate areas as far south as Sydney, Australia.

In the garden, this is an ideal palm for large open areas. Growth is no problem in the tropics. In cooler climates select a warm position free from icy winds,

*Latania loddigesii*
Blue Latan Palm.

*Laccospadix australasica*
This rapidly clustering form makes a good pot-plant for indoors.

*Latania lontaroides*
Red Latan Palm.

*Latania verschaffeltii*
Yellow Latan Palm in fruit.

*Latania loddigesii* x *L. verschaffeltii*
A colourful seed cross.

for example in the protection of a windbreak. Good drainage is essential. Water as needed, especially during drought periods.
**Sex** Dioecious. Need a male and female for seed production.
**Position indoors** No information is available on indoor use.
**Propagation** Seed germinates within 2 months and growth is rapid. Seedlings are reasonably cold-tolerant.

## LAVOIXIA

One species, *L. macrocarpa,* from New Caledonia.
Solitary trunk, with feathered leaves.

## LEOPOLDINIA

Four species, from Amazonian West Brazil. Colombia and Venezuela.
Solitary or clustered trunks, with feathered leaves.

## LEPIDOCARYUM

Nine species, from wet areas of northern South America.
Clustered trunks and leaves in a loose fan shape.

## LEPIDORRHACHIS

One solitary species from Lord Howe Island, growing at high elevations in mossy forests.
*L. mooreana* (syn. *Clinostigma mooreanum*). Little Mountain Palm (Lord Howe Island). A slow grower eventually reaching 2m (6′–7′) high, on a thick ringed trunk almost producing a crownshaft, and deep green feather leaves. The oval red fruits measure 14mm (½″) by 10mm (⅜″).
**Culture and habitat** This species grows at higher altitudes than the *Howea* species, from 800m (2500′) up, where it is accompanied by *Hedyscepe canterburyana.* So far, it is not so commonly grown, and less is therefore known of its characteristics and requirements. Since it originates from higher altitudes, however, it should be hardier than *Howea* in the temperate zones and moister climates, provided there are no frosts. The specimen shown was photographed in the Los Angeles area during one of the city's coldest winters.
**Sex** Monoecious. One plant produces seeds.
**Position indoors** No precise information on indoor use.
**Propagation** Seeds take 6–12 months to germinate, and bottom heat is recommended. Watch for rot in young plants.

# LICUALA

108 solitary or clustered species, occurring from South-East Asia through New Guinea to northern Australia and some of the Pacific Islands, and growing as understorey forest palms. Trunks are single or clumped.

*L. grandis,* Ruffled Fan Palm (New Britain, New Hebrides). The leaves of this most attractive palm resemble a full-pleated skirt of lush green. They are held almost horizontally on a slender trunk to 3m (10′). It makes the most attractive of potted palms, and is very eye-catching. Fruits are crimson and pea-sized 12mm (½″). As an outdoor plant it is suited to the tropics only, but it can be grown in a warm glasshouse in temperate areas.

*L. lauterbachii* (New Guinea). This species has a slender trunk to 4m (12′) high. The almost circular leaves 1m (3′) across are divided into about 30 segments. The flowers are borne in branched sprays 1m (3′) long and the red fruits are 12mm (½″) across. This plant is suited to the tropics only.

*L. paludosa* (Malaysia). This palm has a small solitary trunk. The attractively segmented fan leaves are somewhat similar to those of *L. spinosa,* which is a clumping species.

*L. peltata* subsp. *sumawongii* (India, Burma). This attractive form has large almost circular leaves reminiscent of a partly opened folding fan. The leaves are larger than those of *L. grandis.*

*L. pumila* (Java, Sumatra). This species has a dainty, fan-shaped, pleated leaf, almost forming a full circle.

*L. ramsayi* (syn. *L. muelleri)* (north-eastern Queensland, New Guinea). The circular, pleated leaf of this species splits into a number of segments. The solitary trunk grows to about 6m (20′) or more. With cream flowers and orange-red fruits 1cm (⅜″) across, it is an attractive potted fan palm. It tolerates more cool than the other species, but away from the tropics still requires a warm situation.

*L. spinosa,* Spiny Licuala Palm (Malaysia to Java and Philippines). This clumping palm grows to about 4m (12′) tall. The circular leaves, with spikes on the stalks, are split into about 20 segments and the plant resembles a large clump of *Rhapis.* It prefers shade and requires ample water. The sprays of cream flowers are followed by red fruits 12mm (½″) across. It can be grown in temperate zones if given a warm position.

**Culture and habitat** The Australian species is *L. ramsayi,* and this species is now being grown for indoor use in Queensland, and shipped interstate. Most other species require tropical growing conditions and are rarely seen out of these areas, except in heated glasshouses. They are ideal potted palms.

**In the garden** *Licualas* are mostly tropical rainforest plants and therefore would require to be planted in the protection of trees in suntrap areas. They would need a warm situation in regions outside the tropics, with constant wetting by permanent sprinkler systems. The attractive, almost circular leaves make them well worth a try. Dig in some indoor potting mix over 1–2m (3′–6′) into the soil before planting.

*Latania* sp. (Cairns)
An attractively coloured palm, undoubtedly a seed cross.

*Lepidorrhachis mooreana*
Little Mountain Palm. This specimen photographed at Huntington Gardens, California, much resembles *Howea.*

*Licuala lauterbachii*
**A beautiful palm for the tropics.**

*Licuala paludosa*
**This species resembles *L. spinosa*, but has a solitary trunk.**

*Licuala ramsayi*
**The beautiful Ruffled Fan Palm.**

*Licuala peltata* **subsp.** *sumarwongii*
**The large fan leaves of this palm are magnificent.**

**Sex** Hermaphrodite. Self pollinating for seed production.
**Position indoors** *Licualas* are quite successful and attractive indoor palms. Select a very warm position out of direct sun, a humid sunroom is very good. Hose down the foliage regularly, to maintain humidity and to keep the plant healthy. An occasional application of a foliar fertiliser (such as Aquasol) is also beneficial. Care should be taken not to leave the plants standing in the hot sun afterwards.
**Propagation** Propagation is from seeds, which are best raised with bottom heat, taking 3–6 months to germinate. Some of the clustering varieties may be divided when large enough.

## LINOSPADIX

11 solitary or clustered species, 6 from Australia and 5 from New Guinea; small understorey palms from cooler rainforests.
*L. microcarya* (north-eastern Queensland). This is a small clumping palm 1–2m (3′–6′) high, with about four small, slender dominant trunks, and a number of others remaining as suckers. There is a sparse head of rounded, papery, bright green leaves. A small inflorescence from the base later carries pinkish-red fruits.
*L. minor* (north-eastern Queensland). This small clumping palm has slender trunks up to 1.5m (5′) high, ringed and bare, with a sparse head of leaves

*Licuala ramsayi*
This palm should be grown in shade or undercover.

*Licuala pumila*
This species resembles *L. grandis.*

*Licuala spinosa*
Spiny Licuala Palm.

*Linospadix microcarya*
A small palm suitable for pots or the fernery.

on top. The number of segments per leaf varies from four to 12. The 30cm (1′) inflorescence hangs from amongst the leaves, and flowers are followed by a bunch of small orange fruits that later turn red.

*L. monostachya,* Walking Stick Palm (north-eastern New South Wales and south-eastern Queensland). This is a small palm eventually growing a slender trunk to 3m (10′) high. It has cream flowers and small narrow 18–20mm (¾″) red fruits. The trunk of this species has been harvested for walking sticks.

*L. palmeriana* (north-eastern Queensland). This clumping palm is smaller than the other species, barely reaching 1.5m (5′). The leaves have two lobes forming a V. The inflorescence is small, with cream flowers and then small red fruits. Not pictured.

**Culture and habitat** These slender miniature palms, also known as *Bacularia,* originate from Australia and New Guinea. The species mentioned here may be seen growing in the rainforests from northern NSW to northern Queensland. The most widely grown species is *L. monostachya,* the Walking Stick Palm.

**In the garden** *Linospadix* species require to be sheltered from the sun and wind by taller-growing plants. They also need regular wetting down of the foliage and a moist soil rich in humus and a surface mulch, such as would be found in their natural habitat.

**Sex** Monoecious. Self pollinating for seed production.

**Position indoors** *L. monostachya* has been used for years as an indoor palm in Britain but is rarely seen in Australia, its home. It requires regular wetting down of the foliage, as well as spelling occasionally

*Linospadix minor*
A small, slender clumping palm.

*Linospadix monostachya*
Walking Stick Palm.

in a favourable, regularly hosed position outdoors. This could be in a greenhouse, or under trees (no sun).

**Propagation** Seeds must be fresh and take up to 6 months to germinate. Suckering plants may be divided if the suckers have roots attached.

## LIVISTONA

28 solitary species occurring in north-east Africa, Southern Arabia to the Himalayas and Ryuku Islands and south through Indonesia and Malesia to New Guinea, the Solomons and Australia, which has the most species. The habitat varies considerably, from desert areas with regular water, to mountain rainforests, to swampy flats.

*L. australis,* Australian Fan Palm, Cabbage Tree Palm. Specimens of this hardy palm may be seen dotting the horizon where they have been left in farmland or growing in wet woodland areas along the eastern coast of Australia from north of Brisbane, south to eastern Victoria. They have a ringed, dark trunk to 20m (60′) which is topped with a good head of heavily segmented fan leaves that move freely in the breeze. The spent leaves are shed cleanly after a time. The fruits are reddish brown to black, 12–25mm (½″–1″) across. Juvenile plants resemble *Rhapis.*

*L. carinensis* (syn. *Wissmannia carinensis)* (Somali and Arabia). This species has a trunk 15–20m (45′–60′) high, topped with deep green fan leaves with thorny stalks. The 4–5cm (1½″–2″) fruits are round. A specimen is growing in Brisbane Botanic Gardens, as well as other botanical gardens elsewhere.

*L. chinensis,* Chinese Fan Palm (China). This is the species most grown for the nursery trade. It has a handsome spread of broad fan leaves split into segments only towards the margins. The broad trunk eventually rises to 10m (30′). This species has cream flowers and oblong bluish-grey fruits 12–15mm (½″) long.

*L. decipiens,* Weeping Cabbage Palm, Ribbon Fan Palm (Queensland). Another Australian species from Queensland's central coast, this is somewhat similar to *L. australis,* except that the trunk is not as tall and the fan leaves are longer and more divided and droop lower. Fruits are black and smaller than those of *L. australis.*

*L. drudei* (northern Queensland). On this species the thick trunk is eventually greyish and ringed, to 15m (45′) high. The bright green thin fan leaves have drooping segments, and are divided halfway on long thorny stalks.

*L.* sp. *'Blackdown'* (also known as *L. fulva)* (Blackdown Tablelands, Queensland). As a juvenile this species has stiff, dull grey-green fan leaves, which are brownish beneath. Later maturing dull green above and the leaf segments divided to a third of their length. The 12–15mm (½″–⅝″) round fruits are black with a glaucous bloom.

*L. mariae,* Central Australian Fan Palm. This species is found only in the Macdonnell Ranges near Alice Springs. It differs from many *Livistonas* in colour. It

*Livistona australis*
Australian Fan Palm, Cabbage Tree Palm.

*Livistona australis*
Australian Fan Palm growing in natural habitat in NSW.

*Livistona chinensis*
Chinese Fan Palm, taller palm left rear.

*Livistona carinensis*
Previously *Wissmannia carinensis.*

*Livistona decipiens*
Weeping Cabbage Palm or Ribbon Fan Palm.

*Livistona drudei*
This species resembles *L. decipiens,* but the leaf segments divide only halfway.

has a 15–20m (50–65′) trunk with an open crown of silvery green leaves with reddish stalks. The leaves are also longer and more divided. Creamy yellow flowers are followed by black fruits 15–20mm (½″–¾″) long. The seedlings have a reddish tint.

*L. merrillii* (Philippines). This tall-growing species has a slender trunk topped with a head of spreading and eventually drooping fan leaves with long stalks.

*L. muelleri,* Dwarf Fan Palm (North Queensland). This palm has a solitary trunk, untidy with old leaf bases, to 10m (30′) tall, topped with a dense crown of stiff, segmented fan leaves. The flowers are yellow and the 12mm (½″) blue-black fruits have a whitish surface bloom. It may be seen on the road from Cairns to Port Douglas.

*L. rigida* (Gulf Country of Northern Territory and Queensland). This species has a smooth, ringed trunk to 15m (50′) crowned with an open head of grey-green, heavily segmented fan leaves similar to, but sparser than, those of *L. mariae.* Juvenile leaves are reddish. The hard black round fruits are 12mm (½″) long.

*L. rotundifolia* (Indonesia and Malaysia). In its juvenile stage, this palm has a fan leaf forming almost a full circle and makes a fine tub specimen. Eventually, however, it will reach about 15m (50′) with an attractive head of fan leaves, that becomes more divided. It has yellow flowers and blackish-brown fruits 15–20mm (½″–¾″) long. It is attractive and deserves more popularity.

*L. saribus,* Taraw Palm or Swamp Anaham (South-East Asia). This is a larger species with a clean trunk

*Livistona sp. 'Blackdown'*
Starting to appear in Australian nurseries as *L. fulva*

*Livistona mariae*
Central Australian Fan Palm.

*Livistona merrillii*
A fan palm from the Philippines.
(Photo C. Migliaccio)

*Livistona muelleri*
Dwarf Fan Palm.
(Photo Migliaccio)

*Livistona muelleri*
In natural habitat near Port Douglas, North Queensland.

*Livistona rigida*
The juvenile leaves are reddish.

*Livistona rotundifolia*
Footstool Palm. Has attractive broad leaves.

to 25m (80′) topped by thick head of green fan leaves with reddish stalks armed with vicious spikes. The blue fruits grow in attractively displayed bunches.
*L.* sp. *'Carnarvon'* (Carnarvon Gorge, Queensland). This species has a slender trunk 6–10m (20′–30′) high and has drooping leaf segments. It is somewhat similar to *L. australis* but has yellow flowers. It tolerates sun.
**Culture and habitat** These familiar fan palms from Asia, New Guinea and Australia may be seen in many gardens throughout tropical and temperate Australia. They are also used in landscapes worldwide. The hardiest species is *L. australis* and the most popularly grown is *L. chinensis.* Both of these along with *L. decipiens,* seem to withstand frosts and temperatures almost to freezing point.

As well as garden and street planting and park use, *Livistonas* are widely used as tub or pot specimens, around pools or even indoors. Most species need to be shaded from the hot sun to maintain the lush green colouring, otherwise they will yellow or even burn in patches. A wide veranda provides an ideal situation. For tub specimens, daily watering is essential.

As garden specimens they require room to spread while their foliage is low. Eventually, the trunk raises the foliage up, allowing closer planting beneath. Before planting, dig in some potting mix or humus material for a better start. Water frequently to keep the soil moist.
**Sex** Hermaphrodite, rarely dioecious. Self pollinating for seed production.
**Position indoors** The spread may be a little wide for the average home, but fine specimens may be seen in offices and public buildings where there is good natural light. Under artificial lighting they remain hardy for varying periods. Keep moist.
**Propagation** *Livistonas* seed freely, and the bunches of fruit may be seen hanging amongst the foliage. Seed germinates in about 6 weeks, but early growth is slow.

## LODOICEA

One solitary species, from the Seychelles Islands, where it grows in good restricted stands on the slopes, hills and valleys of two of the islands.
*L. maldivica,* Coco-de-Mer, Double Coconut (Seychelles). A large but slow-growing fan palm, with the trunk reaching up to 25m (80′). The leaves are large, 3–4m (10′–13′) across, with stalks 2–3.5m (6′–12′) long, but they may take 20 years to achieve this size. The flower stems may be 1–2m (3′–6′) long and thick enough to carry 200 kgs (440 lbs) of fruit at maturity. Fruits are the largest in the palm world, about 40cm (16″) long, usually 4 or 5 to a stem, and taking five years to mature. The fruit is like a double coconut joined in the centre, and attains its full size after 12 months, being soft and milky inside. At five years the kernel becomes solid, and its weight can exceed 20 kgs (44 lbs). The plant may take 40 years to bear fruit.
**Culture and habitat** The fertile seeds of this palm

*Livistona saribus*
Taraw Palm of South-East Asia.

*Livistona sp. 'Carnarvon'*
A species from Carnarvon Gorge, Queensland.

*Locoicea maldivica*
Coco-de-Mer or Double Coconut.

*Lytocaryum insigne*
An attractive Brazilian palm, greatly resembling *L. weddellianum*

do not float away to grow on other shores: instead they sink. *Lodoiceas* are not common around the world, in fact they are rarely seen even in botanic gardens. It is possible that many of the leading warmer climate gardens now have specimens started, but because of their slowness it will take many years before they are noticeable. They are difficult to get started and the seeds can take 6–18 months to germinate. The soil needs to be deep to accommodate the germinating seedling's root.
**Sex** Dioecious. Need a male and female for seed production.
**Position indoors** No information available on any usage indoors.
**Propagation** If seeds could be obtained, they would be best planted where they are to grow. They produce a large sinker root which grows to 4m (12′) long and this cannot be disturbed, otherwise the plant will die.

## LOUVELIA

Three species, from East Madagascar.
Solitary trunks and feathered leaves.

## LOXOCOCCUS

One species, *L. rupicola,* from Sri Lanka.
Solitary trunk and stiff feathered leaves.

## LYTOCARYUM

Three solitary species, including palms formerly placed in *Microcoelum, Cocos* or *Syagrus,* from south-eastern Brazil and growing in shady situations in forests at altitudes of 800–1800m (2500′–6000′).
*L. insigne* (Brazil). This is a small palm with a rough trunk to 2m (6′) and displaying a fine array of dark green, shiny leaves with narrow leaflets which are silvery beneath. The inflorescence dangles from amongst the leaves, later bearing oblong cream fruits 2cm (¾″) long. Lovely specimens of this species are one of the most delightful features of the Royal Botanic Gardens, Sydney.
*L. weddellianum,* Baby Cocos, Feather Palm (syn. *L. martianum)* (southern Brazil). This dainty palm eventually grows a slender trunk to about 2m (6′). The fine, glossy green leaves arch downwards, and are whitish beneath. The flowers are yellow and fragrant, and the orange fruits are 20mm (¾″) long and oblong.
**Culture and habitat** These dainty palms are slow-growing and could be best described as resembling a miniature *Cocos plumosa.* In fact, *L. weddellianum* is always sold by the nursery trade under the incorrect name of *Cocos weddelliana.*

In nature, in Brazil, they inhabit the forest floor of high-rainfall forest areas. They are used to low light, and therefore are widely used as indoor palms. They are very adaptable to pot culture and the del-

icate crown of leaves is very attractive. They are commonly known as Baby Cocos Palm and Feather Palm.
**In the garden** They do best under the protection of other plants that give them partial shade from the hot sun. In cooler climates, select a sheltered suntrap position with light cover. Dig in some potting compost or humus. For potted specimens, verandas, screened corners, and protected pool areas are ideal, but do not neglect watering.
**Sex** Monoecious. Self pollinating for seed production.
**Position indoors** These are lovely, dainty palms in the home. Unfortunately, my specimen developed browning of the tips after some time indoors. Maybe it needed more water than I supplied to it. However, moved out onto a veranda, it has grown superbly, and has had no more burnt tips. Perhaps it is best to rotate these palms. Other people I have spoken to have had specimens indoors for three or more years with no problems.
**Propagation** Propagation is from seeds, which germinate in 2–3 months, with bottom heat or other warmth.

## MACKEEA

One species, *M.magnifica,* from north-eastern New Caledonia.

Solitary trunk and feathered leaves.

## MANICARIA

Four species (perhaps one only), from Central America to Trinidad and northern South American regions.

Solitary or clustered trunks and large feathered leaves.

## MAROJEJYA

Two species, from north-eastern Madagascar.

Solitary trunk, with massive leaves partially dividing into feathered segments.

## MASOALA

One species, *M.madagascariensis,* a rare palm from north-eastern Madagascar.

Solitary trunk palm, with feathered leaves.

## MAURITIA

Three species, from north-western South America and Trinidad.

Solitary trunk palm, with fan leaves.

*Lytocaryum weddellianum*
Baby Cocos, Feather Palm. Ideal in pots.

*Mauritiella armata*
Buriti has edible fruits.

*Metroxylon sagu*
Sago Palm is widely used for food.

*Metroxylon salomonense*
Solomon Sago Palm.

## MAURITIELLA

14 or fewer clustered species, from northern South America, growing in lowland areas, in some cases some along the banks of rivers.
*M. armata,* Buriti (Brazil). This species has multiple thorny trunks to 15m (50′) topped by a small head of fully segmented, fan-shaped green leaves which droop when spent. The edible fruits are brown and oblong, 3cm (1″–1¼″) long.
**Culture and habitat** *Mauritiella* species are closely allied to *Mauritia* (not covered in this book). They grow in swampy areas around the Amazon and neighbouring regions. Little is known of them outside botanical gardens. It is thought that they would require damp tropical conditions. In cooler zones select a warm situation, even a suntrap protected from cold winds, and water regularly.
**Sex** Dioecious, sometimes hermaphrodite. Mostly self pollinating to produce seeds.
**Position indoors** No information about their use.
**Propagation** Seeds require warmth and take 2–4 months to germinate.

## MAXBURRETIA

Three species, from southern Thailand and Malaysian Peninsula.
Clustering trunk, shrubby palm, with fan leaves.

## MAXIMILIANA

One species, *M.maripa,* from northern S. America and Trinidad.
Has a solitary trunk and feathered leaves.

## MEDEMIA

One species, *M.argun,* from southern Egypt and northern Sudan.
A solitary trunk fan palm.

## METROXYLON

Five solitary or clustered species, occurring from the Moluccas through to Fiji, and growing thickly in lowland swampy areas. Also widely cultivated throughout South-East Asia for sago. They may have single or multiple trunks, from which the sago is prepared. The trunk is also used for timber and the leaves for thatching.
*M. sagu,* Sago Palm (Moluccas and New Guinea). This is a clumping palm with a thick trunk to 8m (25′) high with numerous suckers at the base. The large 6m (20′) leaves arch upwards and are sometimes prickly, sometimes not. At maturity there is a huge

5–7m (15′–24′) inflorescence, followed by 7–8cm (3″) yellow to brownish fruits, after which the trunk dies.
*M. salomonense,* Solomon Sago Palm (Solomon Islands). This is a clumping palm growing in swampy ground, with dark trunks to 8m (25′) high and developing thorns. The large feather leaves are attractively displayed and have a white scurf at the base of the stalks. Each trunk dies following flowering. Fruits are large 8cm (3″) round and brownish-green.
**Culture and habitat** The sago palms grow in tropical areas of South-East Asia and some of the Pacific Islands. They can only be grown in the tropics (or warm situations in the subtropics) and prefer wet positions.
**Sex** Both monoecious and hermaphrodite (i.e. carrying male, female and bisexual flowers). Self pollinating for seed production.
**Position indoors** No information available.
**Propagation** Seeds require heat to germinate, taking 3–12 months.

## MORATIA

One species, *M.cerifera,* from New Caledonia.
A solitary trunk feather leaf palm.

## MYRIALEPIS

One species, *M.paradoxa,* from South-East Asia and Borneo and Sumatra.
Clustered, climbing rattan, with feather leaves.

## NANNORRHOPS

One clustered species from Iran, Afghanistan, Pakistan and Arabia, located in semidesert areas away from the coast, where water is available not too deep in the soil; it occurs up to an altitude of 1600m (5000′).
*N. ritchiana,* Mazari Palm (Iran, Afghanistan, Pakistan, Arabia). This unarmed palm with clustered and branching trunks has blue-green fan leaves with upright stalks. The branched inflorescence stands high above the leaves. The round orange-brown fruits are 13mm (½″) in diameter.
**Culture and habitat** *Nannorrhops* is a slow-growing palm, hardy over a wide range of temperatures and also tolerant of dry conditions. It will withstand biting winds and moderate snowstorms. As is the case with *Corypha* and many similar palms, each stem dies after flowering and seed production and is replaced by a new growth.

In its native countries this palm is a source of green 'cabbage', and the fruits are also eaten. However, its main use is to provide fibre for rope making and weaving.

If you want to grow this palm you should give consideration to the conditions in its natural habitat avoiding excesses of moisture and humidity.

*Nannorrhops ritchiana*
The Mazari Palm is desert-hardy.

*Neodypsis decaryi*
Madagascan Three-sided Palm or Triangular Palm.

*Neodypsis lastelliana*
A less commonly seen species.

*Neonicholsonia watsonii*
This rare palm was only recently relocated in nature.

**Sex** Hermaphrodite. Self pollinating for seed production.
**Position indoors** Not known.
**Propagation** No precise information available on germination times.

## NENGA

Five species, from Vietnam south to and including Indonesia.

Tall solitary trunk feather palms.

## NEODYPSIS

14 solitary or clustered species from Madagascar, growing in tropical and mountainous rainforests from sea-level to 2000m (6000′).
*N. decaryi,* Madagascan Three-sided Palm, Triangular Palm (Madagascar). This species is both rare and an oddity, with the foliage emerging from the 3–5m (10′–16′) trunk in a triangular arrangement of leaf bases.

The handsome, grey-green feathered leaves have short brown stalks with a whitish scurf, and arch gracefully. The inflorescences appear from among the lower leaves and later bear over 2.5cm (1″) fruits. Now being grown commercially in Brisbane, Australia, this outstanding and spectacular palm should be heavily sought after. It was used on the World Expo site at Brisbane, Queensland.
*N. lastelliana* (Madagascar). This is a handsome medium-sized feather palm, still quite rare in cultivation, and the crownshaft and stalks of the erect fronds are attractively covered in a red fuzz. The stiffish green leaflets stand out at right angles to the erect stalks.
**Culture and habitat** Of the 14 species of *Neodypsis, N. decaryi* is the most interesting, with its triangular growth pattern. Outdoor specimens are growing quite well in both Cairns and Brisbane botanic gardens, indicating its tolerance of both tropical and subtropical climates. *N. lastelliana* is grown in the Los Angeles and survives the coldest of winters in that warm temperate area.

Both species require full sun and will withstand a drier position once established. In warm temperate localities they should be planted in a suntrap situation protected from cold and cold winds.
**Sex** Monoecious. Self pollinating for seed production.
**Position indoors** No information available on indoor use.
**Propagation** Seeds germinate within 1 month.

## NEONICHOLSONIA

One variable mostly solitary species, from Nicaragua to Panama, growing in low altitude from sea-level to 250m (800′).

*N. watsonii* (Costa Rica, Panama). A rare palm closely allied to *Euterpe.* It grows to 2–3m (6′–10′) high and has 1m (3′) green feathery fronds, the leaflets of which taper to a long point. The inflorescence is a spike and the 11mm (½″) black fruits are oval and show scars.
**Culture and habitat** This rare palm was lost to culture for some time, but has since been rediscovered. Its habitat is in the shade of tropical rainforests in Costa Rica and Panama, where it grows with *Bactris* and *Euterpe.* It requires a tropical or warm subtropical climate and needs similar growing conditions to *Euterpe.* In nature this palm is only to be found in rainforests, so it would be best planted amongst other palms or trees. Adding a surface mulch would help, and it should receive regular water. It could also be planted in a bush house, being a smallish grower, or grown in pots.
**Sex** Monoecious. Self pollinating for seed production.
**Position indoors** This palm is similar to *Euterpe,* which is used indoors, so it may be worth trying indoors.
**Propagation** Should be similar to *Euterpe.* No germination times are available.

## NEOPHLOGA

29 solitary or clustered species from Madagascar, located in both lowland and mountain rainforests.
*N. scottiana* (Madagascar). An upright-growing, slender palm with a solitary, brownish-green ringed trunk. The feather leaves are held out horizontally from the trunk, and the leaflets are narrow and blue-green, while the new leaf bases are glaucous silver.
**Culture and habitat** Neophlogas are the *Chamaedoreas* of Madagascar. They are small understorey palms. Most, resemble species of *Chamaedorea* and should be grown under similar conditions. As these palms are still little known around the world, few specimens exist outside Madagascar.

They can be grown in a humid greenhouse, either in the ground or in pots. Alternatively, you may find that, like *Chamaedoreas,* they are happy in less pampered conditions away from direct sun. We will not know much about cultivation of *Neophloga* species until stocks become more plentiful. Refer to *Chamaedorea* for further information.
**Sex** Monoecious. Self pollinating for seed production.
**Position indoors** No information available.
**Propagation** No precise information available on germination times.

## NEOVEITCHIA

One solitary species from Viti Levu in Fiji, located in the bottom of one valley and in danger of extinction because of land clearing for fruit plantations in response to increasing population.
*N. storckii,* Unleito (Fiji). This palm has a heavier trunk than *Veitchia (Neoveitchia* means 'new *Veitchia'*), up

*Neophloga scottiana*
An understorey palm from Madagascar.

*Neoveitchia storckii*
Unleito, from Fiji, is endangered.

*Nephrosperma vanhoutteanum*
Latanier Millepattes.

*Normanbya normanbyi*
Black Palm with *Ptychosperma elegans* at the rear.

to 15m (50′) tall and lacking a crownshaft. The feather leaves are heavier and more numerous than those of *Veitchia.* The reddish-brown fruits are oval, 3cm (1¼″) by 2cm (¾″) and are eaten by the local people.
**Culture and habitat** *Neoveitchia* grows alongside *Veitchia joannis* in its natural habitat and its growing conditions are similar. Refer to *Veitchia* for cultural details.
**Sex** Monoecious. Self pollinating for seed production.
**Position indoors** Not known.
**Propagation** No precise information is available on seed germination times.

## NEPHROSPERMA

One solitary species from The Seychelles, growing in lowland light forests and on rocky slopes below 500m (1500′).
*N. vanhoutteanum,* Latanier Millepattes (Seychelles). This slender palm has a trunk 6–10m (20′–30′) high topped with a cylinder of glossy green leaf bases resembling a crownshaft. The leaf fronds are feathery, 1–1.5m (3′–5′) long, and arching. The 1.5–2m (5′–7′) inflorescences, longer than the leaves, come from below the crownshaft, and later bear 15mm (½″) red kidney-shaped fruits. On younger plants the leaf stalks may have black spines.
**Culture and habitat** A shade-loving palm for the tropics and subtropics. It has been cultured for many years by collectors and is a favourite in glasshouses as well as in warmer botanic gardens. It appears to grow in the tropics and subtropics without any special requirements other than ample water, good drainage, reasonably good soil and shelter from cold wind.
**Sex** Monoecious. Self pollinating for seed production.
**Position indoors** No information available on indoor use.
**Propagation** Seeds will germinate within a few weeks. This palm is wind- and cold-tender (young plants especially so), and therefore always needs shelter from the elements.

## NORMANBYA

One solitary species from North Queensland, growing in high-rainfall vine-forests in gravelly soils along rivers and streams.
*N. normanbyi,* Black Palm (North Queensland). This palm has a tall slender trunk to 20m (60′) high which becomes almost black. There is a pale crownshaft topped with a dainty head of leaves, the leaflets of which appear to be cut off at the ends, and in maturity become densely plumed. The flowers are borne in a clustered bunch below the crownshaft. The fruits are about 5cm (2″) long, pear-shaped and pinkish to brown.
**Culture and habitat** *Normanbya* is a handsome palm when mature. It comes from the tropical rainforests of northern Queensland but can be adapted to warm

positions in zones from tropical to temperate. It should be grown more than it is, but remains relatively unknown outside small circles of enthusiasts. The plumed head of leaves is most attractive and eye-catching.

As a rainforest dweller, this palm would be best planted in the shelter and shade of other trees or palms. As it grows taller, it will emerge out into the sun. *Normanbya* requires plenty of moisture and relishes constant wetting down of the foliage. Dig in some potting mix or compost before planting. Avoid strong coastal or dry inland winds. In temperate areas, select a sheltered suntrap position away from cold winds.

**Sex** Monoecious. Self pollinating for seed production.

**Position indoors** Only small use is made of *Normanbya* indoors, so little information is available. Maybe this lack of indoor use is because in its younger stages it is tall to 2m (6′) in a 200mm (8″) pot) with only a sparse and relatively unattractive head of cut off leaves. Later it takes on its mature form with the beautiful plume-like leaves, but only at a size that would require a larger pot and a high ceiling. Good lighting, preferably near a skylight or in a sunroom, is desirable. Water well and feed with a complete plant food occasionally.

**Propagation** Seeds are scarce and, if fresh, germinate quickly (within 3 months).

*Normanbya normanbyi*
There is a fine-leaf species, as shown in this juvenile specimen.

## NYPA

One clumping species, from the Ganges delta and Sri Lanka through Malaysia, Indonesia, New Guinea and northern Australia to the Solomon and Ryukyu Islands. It grows with mangroves in the soft mud of river estuaries.

*N. fruticans,* Mangrove Palm (South-East Asia, Australia and Pacific Islands). This palm grows in brackish water, and has a clumping trunk which is usually submerged in mud. The feather leaves are long, deep green above and whitish beneath. The yellow flowers are borne on a 1–2m (3′–6′) spike from the base of the plant and the 10–15cm (4″–6″) brown fruits, carried in a club-like head, are edible.

**Culture and habitat** This aquatic palm grows in muddy streams and coastal watercourses, usually brackish, through coastal South-East Asia and neighbouring tropical areas. In these countries its immature fruits are eaten along with its sugary sap content, tapped for drinks and alcohol. Its foliage is used for thatching.

Outside its natural habitat, *Nypa* is mainly grown in botanic gardens, which can usually duplicate its environmental conditions better than a home grower can. However, I have noticed that *Nypa* has often been planted in aspects where the soil is badly drained but not permanently waterlogged, and this widens its scope for inclusion by private enthusiasts in their plantings.

**Sex** Monoecious. Self pollinating for seed production.

**Position indoors** No information available.

*Nypa fruticans*
Mangrove Palm has edible fruit.

*Oenocarpus* sp. (juvenile)
A clump-forming palm. Oil is obtained from the fruit.

*Oncosperma tigillarium*
Nibung Palm forms a clump.

**Propagation** Seeds germinate best in brackish water. In nature, seeds often germinate after being dispersed by water. No information is available on germination times.

## OENOCARPUS

Eight solitary or clumping species, occurring from Costa Rica to Bolivia and Brazil, and growing in sandy soil in rainforests and on firm river banks. *Oenocarpus* species come from the tropical areas of Central and South America. There are both single- and multiple-trunked species. The trunk is topped by feather leaves which have narrow, widely spaced leaflets. The trunks have the appearance of bamboo, with their widely spaced rings and no distinct crownshaft. The short bushy inflorescences hang from below the leaves, later bearing dark purple oval fruits, 2–3cm (¾″–1⅛″) long.

*O.* sp. (most likely *O. panamanus* from Panama).
This species has multiple trunks and feather leaves with chestnut bases and widely spaced deep green leaflets, as a juvenile. With maturity, these become closer and the tips gracefully droop.

**Culture and habitat** *Oenocarpus* species come from dry tropical areas where it is hot at all times. If available, they should grow in tropical and subtropical areas without much trouble.
*Oenocarpus* do not seem to have been grown widely, and few mature specimens seem to exist even in any of the botanic gardens; knowledge of them is scanty. The richly coloured specimen photographed was raised at the Royal Botanic Gardens, Sydney, and the species has not yet been identified. No doubt it will soon be planted out.
Should you obtain an *Oenocarpus,* it will require a well-drained or even sandy soil and a warm position sheltered from cold winds and preferably with overhead cover but not too dense,so that the warmth can get through. Tropical or subtropical localities would be ideal, but the specimen photographed was growing in Sydney, which is in the warm temperate zone.

**Sex** Monoecious. Self pollinating for seed production.

**Position indoors** No information available.

**Propagation** It is reported that seeds germinate within 2 months. Seeds would probably require bottom heat and young plants would probably need glasshouse conditions.

## ONCOCALAMUS

Five species, from equatorial West Africa.
Clustered, climbing rattans, with feather leaves.

## ONCOSPERMA

Five clustered species, spread through South-East Asia, from Sri Lanka and Indonesia to the Philippines,

growing in habitats ranging from mangrove swamps to forests at altitudes of 300–1000m (1000′–3000′).
*O. tigillarium,* Nibung Palm (South-East Asia). A multitrunked feather palm to 25m (80′) forming a large clump. The trunks, which are covered in black spines, are topped with dark green feather leaves on stalks which are also armed. The armed inflorescences produce purplish black roundish fruits, broader than they are long, 12mm (½″) by 10mm (⅜″).
**Culture and habitat** *Oncosperma* species form very large clumps in the coastal swampy areas in South-East Asia. Because of their eventual size and their spines they are best confined to parks in tropical areas.

These tropical palms require high humidity and are very sensitive to cold. They would be best grown in the tropics, but if you are attempting to grow them in the subtropics, they should be confined to warm suntrap positions where humidity can also be maintained. Because they come from swampy areas, regular watering is also essential. Being spiky, they should be placed away from traffic areas, where they might be accidentally walked into.
**Sex** Monoecious. Self pollinating for seed production.
**Position indoors** No information available.
**Propagation** Seeds must be fresh and then germinate with warmth within 2 months.

*Orania palindan*
*Orania* palms are poisonous.

## ORANIA

17 solitary species from eastern Madagascar, Indonesia, Malaysia, the Philippines and New Guinea.
*O. palindan* (Philippines). This species has an attractive 5–7m (15′–23′) ringed trunk bearing upright 2–4m (6′–12′) feather leaves with the leaflets ragged at the edges. The drooping inflorescence of cream flowers matures to bear 4–5cm (1½″–2″) orange-green to brownish pear-shaped fruits.
**Culture and habitat** *Orania* species are tall canopy palms of the tropical rainforests, mostly from the lowlands to 1700m (5500′); a few varieties are subcanopy.

They would require tropical to subtropical conditions with regular watering. It is doubtful whether they would be successful in climates cooler than coastal areas of latitudes 30° N and S, other than under glasshouse protection.

*Oranias* grow in areas of high rainfall, with the foliage more often wet than dry. The soil is usually rich, but well drained. They would do best growing amongst other palms or moderately dense vegetation such as a rainforest supplies. Keep regularly watered and away from sharp ocean winds.

Most *Oranias* are poisonous, which probably accounts for their lack of popularity, considering their handsome appearance. Care should be taken to remove the poisonous fruits, so that children won't eat them.
**Sex** Monoecious. Self pollinating for seed production.
**Position indoors** No information is available.

*Oraniopsis appendiculata*
A handsome Australian palm (prev. listed as *Orania*)

*Oraniopsis appendiculata*
Is closely related to *Ceroxylon*, the wax palm.

*Orbignya cohune*
*Cohune* Palm yields palm oil.

**Propagation** Seeds germinate within 3 months and bottom heat is desirable. They produce a very long taproot, and should therefore be potted up into long containers early. For home growers they would be best planted where they are to grow.
*Oraniopsis appendiculata* was formerly placed in this genus also.

## ORANIOPSIS

One solitary species, from north-eastern Australia, growing at elevations of 300–1500m (1000′–5000′) in mountainous rainforest and gullies with high rainfall.
*O. appendiculata* (north-eastern Queensland). This palm has a solitary grey ringed trunk 6–10m (20′–30′) high topped by a head of feather leaves which are dark green above and grey beneath. The inflorescence, which appears from amongst the leaves, has yellowish flowers followed by 3cm (1⅛″) long orange-yellow fruits.
**Culture and habitat** This palm grows in mountain rainforests north from Innisfail.

It was originally known as *Orania appendiculata*, but following research, linked with the wax palms *Ceroxylon* of South America.

In the tropics and subtropics plant in warm, frost-free, protected gardens which receive an abundance of moisture. In cooler areas select a protected suntrap position, amongst other palms or plants for protection against cold winds. The soil should be rich and well drained, and a surface mulching will help to maintain humidity. In the natural habitat of this palm, the foliage is regularly moistened by mountain mists and regular rainfall. This should be duplicated to retain the beautiful appearance of the foliage. It is one of the most handsome palms grown.
**Sex** Monoecious. Self pollinating for seed production.
**Position indoors** Little is known, but the handsome foliage makes it worth a try. A sunroom might be best. It may also be used as a glasshouse specimen in cooler areas.
**Propagation** Propagation is from seeds, which germinate best with bottom heat, taking 3 months. The shoot goes straight down deeply, before leaves appear above the ground. Use deep pots, or plant the germinating seeds directly into the garden.

## ORBIGNYA

20 solitary species, occurring in Central America and northern South America, growing under varying conditions, from humid tropical rainforests to open savanna or areas which flood.
*O. cohune*, Cohune Palm (Central America). This palm has a mass of 10m (30′) long feather leaves, which stand upright and are almost stalkless. The 25cm (10″) wide trunk eventually grows to a height of 15–20m (50′–60′). The dead leaves and old bases often adhere to this grey scarred trunk. The inflor-

escences hang from the lower leaves and the oval fruits 6–7cm (2 ½″) long, resemble small coconuts.
*O. lydiae* (Brazil). This species is similar to *O. cohune* and is on trial at the Royal Botanic Gardens, Sydney. Not pictured.
**Culture and habitat** *Orbignya, Attalea* and *Scheelea* are all very similar in appearance and collectively account for about 80 species of palms. In their native regions their large, broad leaves are used for thatching and the leaflets for weaving. The seeds are very rich in oil which, while inferior to that of the commercial oil palms, is used in soap making. *Orbignyas* prefer a tropical climate, but will also grow quite well in the subtropics, and may be grown as single or massed specimens. They are best displayed on either side of a walk or driveway, where their arching leaves form a covered tunnel.

These are very large palms, requiring a lot of space, and best suited to parks or large gardens. Avoid cold winds in the cooler zones and maintain ample and regular watering, to maintain humidity.
**Sex** Monoecious. Self pollinating for seed production.
**Position indoors** Little is known of their indoor use. Try a shaded suntrap corner, near a window, or a warm shaded sunroom corner.
**Propagation** Fresh seeds germinate within 2 months, but growth is slow.

## PALANDRA

One species, *P. aequatorialis,* from Ecuador.
Solitary trunk feather palm.

## PARAJUBAEA

Two species, from Ecuador, Bolivia and Colombia.
Solitary trunk feather palm.

## PELAGODOXA

One solitary species, from the Marquesas Islands, and rare, growing in dense rainforests in humid valleys from sea-level to 135m (450′).
*P. henryana,* Vahana Palm (Marquesas Islands). An unusual and rarely grown palm, but a collector's gem. The trunk will attain a height of 6–7m (20′–24′) and is topped with a nicely displayed head of leaves which have yellowish stalks, tending to have a white coating lower down. The leaf blades are almost oval, 2m (6′) x 1m (3′), showing grooves but remaining unsplit except at the tips, unless damaged by the wind. The flowers are borne amongst the leaves, and the large warty and corky round fruits are about 6–10cm (2½″–4″).

**Culture and habitat** This decorative palm is much sought after as an ornamental. Being near-equatorial in origin, it requires a tropical climate and sheltered conditions to look good. It will probably never be

*Pelagodoxa henryana*
Vahani Palm, the large leaves are undivided.

*Pelagodoxa henryana*
Looking up from under a large specimen.

*Phoenicophorium borsigianum*
Latanier Feuille. A beautiful palm.

*Phoenix acaulis*
A species from India and Burma.

offered commercially because of scarcity of seed stocks. Away from the tropics it would be best grown in a glasshouse as was the specimen photographed in the tropical dome at Brisbane Botanic Gardens.

This palm has rather restrictive requirements and, because of very limited seed production and natural stock, may never be available on the open market. However, should you have a plant, it would require a warm, shady, protected position in a tropical climate. It would be important to provide humidity by using a surface mulch and maintaining regular watering. This plant looks best if grown away from the wind.

**Sex** Monoecious. Self pollinating for seed production.

**Position indoors** No information.

**Propagation** Seeds are difficult and erratic to germinate. Young plants would need protection. No germination times are available.

## PHLOGA

Two species, from Madagascar.

Solitary or clustered trunk, with feather leaves.

## PHOENICOPHORIUM

One solitary species, also called *Stevensonia,* from The Seychelles, growing in the humid valleys and on slopes, from sea-level to 300m (1000′).

*P. borsigianum,* Latanier Feuille (Seychelles). In its native habitat this palm has a slender 15m (50′) trunk, spiny at first, later smooth and ringed. The broad feather leaves are very handsome and are only split at the ends. The inflorescence is borne amongst the lowest leaves and is followed by 12mm (½″) greenish-yellow fruits which ripen reddish.

**Culture and habitat** This tropical, warmth- and moisture-loving palm grows on slopes and in valleys of the Seychelle Islands, to an elevation of about 300m (1000′). It is very handsome and is much admired, but requires such exacting conditions for its survival that it is rarely grown outside the tropics except in heated glasshouse. Like *Pelagodoxa,* this palm needs a tropical climate, a humid position free from wind, and plenty of regular water to prevent tip burn. Surface mulch would be a further asset. This is also an attractive potted palm for heated glasshouses or other suitable situations.

**Sex** Monoecious. Self pollinating for seed production.

**Position indoors** No information.

**Propagation** Seeds require bottom heat to germinate, and seedlings require heated glasshouse conditions. No germination times are available.

## PHOENIX

17 solitary or clustered species, occurring in southern Madagascar and Africa, through the Middle East and

Asia to South China and the Philippines. They are mostly confined to arid areas with available water, but there are a few monsoonal varieties also.
*P. acaulis* (Bengal to Burma). Usually trunkless, this species sometimes does develop unattractive trunks, which are generally littered with old leaves on their bases. The feathery leaves are dark green, about 2m (6′) long and stiff, but nicely displayed. The inflorescence is 40–60cm (16″–24″) long and the fruits are bright red.
*P. canariensis,* Canary Island Date Palm (Canary Islands). This palm has a heavy trunk up to 20m (60′) high (covered with old leaf bases). There is a large head of spreading, heavy-stalked leaves, to 6m (20′) long and sharply spiked at the base. It requires plenty of room to spread. The orange 2cm (¾″) fruits are relished by birds.
*P. dactylifera,* Date Palm (western Asia and northern Africa). The commercial Date Palm has a thick, patterned trunk to 20m (60′) high with suckers at the base. On outstanding varieties these are removed and grown. There is a sparse head of 6m (20′) long, grey-green leaves, with the spent leaves hanging dead below. Bunches of oblong orange fruits to 6cm (2½″) form only in hot, dry climates. The later commercially developed varieties have much larger fruits.
*P. loureirii* (India to southern China). This is a clump-forming palm, somewhat similar to *P. pusilla,* but growing taller and having fewer trunks. The stiff, dark green leaves have leaflets arising in different planes, giving a full and plumose appearance. It has cream

*Phoenix canariensis*
Canary Island Date Palm.

*Phoenix dactylifera*
A single-trunked form.

*Phoenix dactylifera*
A clumping form of the commercial Date Palm.

*Phoenix loureirii*
**This plant is growing in Brisbane Botanic Gardens.**

*Phoenix pusilla*
**Ceylon Date Palm.**

*Phoenix reclinata*
**Senegal Date Palm.**

*Phoenix roebelenii*
**Dwarf Date Palm with fruits.**

flowers and reddish to purplish black fruits. Old leaf-bases persist.

*P. pusilla* (syn. *P. zeylanica),* Ceylon Date Palm (India and Sri Lanka). This solitary or clumping palm has short trunks of 2–3m (6′–10′) which are said to yield a meal or flour, used in times of scarcity. The pale green 2m (6′) leaves are very spiky. Flowers are borne amongst the foliage and are followed by blackish purple fruits which are sweet and edible. This species grows on seashores.

*P. reclinata,* Senegal Date Palm (tropical Africa and Madagascar). This date palm has clustering, slender trunks to 15m (50′) high that recline away from one another so that the heads of 2–3m (6′–10′) deep green leaves are separately displayed. Fruits are orange to reddish and oval, to 2cm (¾″) long.

*P. roebelenii,* Dwarf Date Palm (India to China). Most widely grown for its dainty juvenile form, this species eventually grows a trunk to 3m (10′) high and leaves to 1.5m (5′) long which recurve downwards. Fruits are oblong and black, 12mm (½″) long.

*P. rupicola,* Cliff Date Palm (India). This palm has a smooth trunk to 8m (25′) high. The 3m (10′) long bright green leaves have a flat appearance. The 2cm (¾″) long fruits are shiny yellow to reddish when ripe.

*P. sylvestris,* Silver Date Palm (India). This species, resembling *P. canariensis* but with silver-green leaves, has a thick, patterned trunk to 10–15m (30′–50′) high. It carries spikes of white flowers, followed by orange-yellow fruits. The sap yields commercial quantities of sugar.

**Culture and habitat** The date palms are probably the most widely cultivated of all palms. They are reasonably hardy, being seen in most countries of the world, except in icy regions. In nature, some of the species grow in hot semiarid regions, while others come from more humid areas.

The larger species may be grown as garden or park specimens or used for street plantings, while the small *P. roebelenii,* Dwarf Date, is more often used in pot culture. All species can be used in this role indoors until they become too large or spiky. Then they may be planted outdoors in the garden or, in the case of the smaller varieties, repotted in a larger container for continued outdoor pot culture. The older, spent leaves, always need to be removed to keep the plant tidy.

*Phoenix* species fruit regularly and their fruits are enjoyed by the birds. When the fruit is ripe a *P. canariensis* will often be seen swarming with currawongs and other larger birds. The fruit of *P. dactylifera* is our much-loved commercial date.

**In the garden** Select the position to suit the variety to be grown. Do not jam a wide, spreading *P. canariensis* into a corner where its spines will stab every passerby. Give it room to spread and be seen to best advantage. *Phoenix* species are quite hardy, even in the poorest of soils or the driest of positions. However, the better their positions the healthier their appearances. Regular watering will also give improved results.

**In pots** *Phoenix* are good potted specimen palms, both indoors and outdoors. Being evenly shaped,

*Phoenix rupicola*
Cliff Date Palm.

*Phoenix sylvestris*
Silver Date Palm.

*Pholidocarpus mucronatus* (juvenile)
Kepaw is used for thatch.

*Pigafetta filaris*
Wanga Palm requires warmth.

they are attractive in all pot sizes, from 100mm (4″) to 500mm (20″). Pot on as required, using an indoor potting mix. Feed occasionally with a complete plant food to maintain a nice dark green colour and lush growth.
**Sex** Dioecious. Needs a male and a female for seed production.
**Position indoors** All *Phoenix* species may be used indoors, although some can be quite large and spiky, and dangerous if bumped into. The most used is *P. roebelenii.* Select a bright position for longest indoor life. Take outdoors again at first signs of distress. Keep moist, but not excessively wet, nor overly dry. Also, observe for mealy bug infestation indoors. Take outdoors and hose down occasionally, or rotate plants regularly with veranda specimens.
**Propagation** Seeds germinate in 2–3 months. Suckering varieties may also be propagated from suckers.

## PHOLIDOCARPUS

Six solitary species from southern Thailand to Indonesia, growing in lowland swamp and waterlogged areas.
*P. mucronatus,* Kepaw (Sumatra). A tall fan palm with a rough, ringed trunk to 15m (50′) high and 30cm (12″) thick. The crown of armed fan leaves resembles that of a *Livistona,* the leaves being deep green and stiffly held. The inflorescence, from amongst the leaves, is long and branched with cream flowers followed by rather large 8cm (3″) brown, scaly fruits.
**Culture and habitat** *Pholidocarpus* species are tall fan palms similar in stature to *Livistonas* and *Washingtonias.* They grow in coastal areas near the sea or in swampy forest areas and are rarely grown outside their natural habitats. This is probably because of the widespread use of the other similar palms.

For anyone having stocks and wishing to plant *Pholidocarpus,* they should be easy enough to grow in moist tropical areas and coastal subtropical areas away from cold winds and frosts. Plant in the same way as *Livistona,* making sure they are kept moist.
**Sex** Hermaphrodite. Self pollinating for seed production.
**Position indoors** No information available on indoor use.
**Propagation** Seeds germinate best with bottom heat.

## PHOLIDOSTACHYS

Three species, from Costa Rica to Peru.
Solitary trunk feather palms.

## PHYSOKENTIA

Seven species, from Fiji, New Hebrides, Solomon and New Britain Islands.
Solitary trunk feather palms.

## PHYTELEPHAS

12 species, from regions of north-western South America, to Panama.

Solitary or clustered trunk palms with feather leaves.

## PIGAFETTA

One solitary species from eastern Indonesia and western New Guinea growing on mountainsides, old landslides and lava flows, or river banks, from 300–1500m (1000′–5000′) and preferring high light intensity.

*P. filaris,* Wanga Palm (Indonesia, New Guinea). A tall, sun-loving feather palm with the juvenile stage rather prickly, but rapidly growing on to develop an attractive grey-ringed green trunk to 40m (120′) high. A rapid-growing, tall, straight palm, with the inflorescence coming from amongst the lower leaves and bearing oval, scaly white and mauve fruits 12mm (½″) long.

**Culture and habitat** *Pigafetta* species are tall and handsome palms with a liking for space rather than crowded rainforests. They flourish out in open areas, and can quickly establish on freshly cleared ground. In their native countries their straight trunks are used for timber. They also require very rich soil and very high rainfall. While they do grow in rainforest areas,

*Pinanga barnesii*
A showy small clumping palm with similar foliage to *P. maculata.*

*Pinanga elmeri*
This species forms a dainty clump.

*Pinanga coronata*
An elegant cluster palm for the tropics.

*Pinanga insignis* (juvenile)
A tall single-trunked species with dark leaf stalks.

seedlings cannot establish in such heavy shade, preferring open cleared areas. They will not tolerate cold, so are best confined to the tropics. Nevertheless, the specimen photographed was growing happily in the subtropical climate of Brisbane.
**Sex** Dioecious. Needs a male and a female for seed production.
**Position indoors** No information on use indoors.
**Propagation** Fresh seeds are said to germinate quickly, within 1 month, and require an open sunny position. Seedlings may die if shaded.

## PINANGA

120 solitary or clustered species, occurring in South-East Asia from the Himalayas to southern China and New Guinea, growing in a variety of soils in rainforests, from sea-level to an altitude of 2800m (9000′). The common names of these palms are Pinang and Bunga.
*P. barnesii* (Philippines). This is a small clumping palm with slender purplish trunks to 2m (6′) high topped with broadly segmented green leaves which are marbled with purple. Fruits are oval, bright red, 2.5–3.5cm (1″–1½″) long. An attractive small palm for shaded warm and tropical areas.
*P. coronata* (Java). An attractive cluster palm with slender golden trunks to 4m (12′) topped with a dainty head of bright green feather leaves. The oval 12mm (½″) fruits are red. This is supposed to be the correct

*Pinanga kuhlii*
Ivory Cane Palm.

*Pinanga maculata*
Tiger Palm has a single trunk.

name for *P. kuhlii,* but specimens identified as the latter, that I have seen, appeared entirely different in leaf shape.

*P. elmeri* (Philippines). This is a graceful species forming an attractive clump. The feather leaves with yellow stalks are attractively displayed, and the leaf bases have rusty brown scales.

*P. insignis* (Philippines). This has a larger trunk than other species, 7–9m (20′–30′) high, topped with deep green feather leaves. Fruits are red.

*P. javana* (West Java) This 10m (30′) solitary trunk variety has a prominent crownshaft and bright green feather leaves which are nicely formed. The fruits are small.

*P. kuhlii,* Ivory Cane Palm (Indonesia). This attractive clustering feather palm has slender yellow trunks topped with broadly segmented bright green leaves. Pinkish-cream flowers are followed by oval dark red fruits (12mm) (½″) long. This species may be synonymous with *P. coronata.*

*P. maculata,* Tiger Palm (Philippines). This species is similar to *P. barnesii* and is often mistaken for it, but has a single trunk. It is a most attractive palm.

*P. merrillii* (Philippines). This interesting palm with multiple pinkish trunks to 4m (12′) high is growing in Brisbane Botanic Gardens. It prefers shade and tropical conditions. This species comes from lowland areas, while a very similar species, *P. elmeri,* is found in the highlands.

*P. minahassai* (Sulawesi). This is a colourful species

*Pinanga merrillii*
A species with dark leaf stalks.

*Pinanga minahassai*
Under trial at Sydney Botanical Gardens.

*Pinanga modesta*
A juvenile specimen under trial.

*Pinanga patula*
This palm forms a tall clump with green crownshafts.

*Pinanga scortechinii*
Another tall clumping species, with golden crownshafts.

with the deep green feather leaves shaded purple on purple stalks; the trunk is also purple.
*P. modesta* (Philippines). A recent acquisition at the Royal Botanic Gardens, Sydney. The juvenile plant will need further study.
*P. patula* (Sumatra). The bamboo-like trunks of this species grow to about 4m (12′) and are topped with a crownshaft turning yellowish as it becomes leaf bases. There are about five to eight feathered leaves 1m (3′) long. The inflorescence is small and branched.
*P. scortechinii* (India to Malaysia). This species has a cluster of bamboo-like trunks to 4m (12′) high topped with a long yellow crownshaft. The 1m (3′) feather leaves are borne on 1m (3′) stalks. The inflorescence dangles from below the crownshaft.
*P.* sp., Ivory Crownshaft (northern Sumatra). Multiple cream trunks to 5m (15′) are topped with a handsome display of bright green feather leaves. The red fruits are oval and 12mm (½″) long.
**Culture and habitat** *Pinangas* are tropical palms from rainforests and other shady habitats from Indonesia to the Philippines. They are becoming very popular with palm enthusiasts in the tropics (where they grow effortlessly), because of their attractive appearance, medium size and colourful fruit display. There are many species not yet named, which is a shame, but no doubt these will be sorted out in the future.

Many of the *Pinanga* species, besides doing well in the tropics, seems to have adapted quite well in the subtropics and warm temperate areas. Here they should be given a warm but shaded position and

*Pinanga 'Ivory Crownshaft'*
A species with handsome feather leaves, grown under the name of Ivory Crownshaft in Cairns.

protected from strong, cool winds. In cooler areas they would have to be treated as glasshouse specimens, either in pots or planted in garden beds.
**In the garden** Select a warm, shaded position with protection from salty, cool or strong winds. A warm rainforest-type mass planting would be ideal. So, it should be well drained, with some compost or potting mix added to raise the level, and a regular water supply must be maintained.
**In pots** *Pinangas* are wonderful potted specimens. Pot up when necessary, using an indoor potting mix. Water regularly and observe the above shade and protection requirements.
**Sex** Monoecious. Self pollinating for seed production.
**Position indoors** Select a warm position which receives good natural light. Sunrooms are ideal. The plants should be kept moist, but must not stand in saucers of stagnant water.
**Propagation** Seeds germinate better with bottom heat, taking about 2 months. Protect young plants outside the tropics.

## PLECTOCOMIA

16 species, from the Himalayas and South-East Asia to the Philippines.
Solitary or clustered climbing rattans with feather leaves.

## PLECTOCOMIOPSIS

Five species, from southern Thailand, Malay Peninsula, Borneo and Sumatra.
Clustered, climbing rattans, with feather leaves.

## PODOCOCCUS

One species, *P. barteri,* from West Africa.
Small solitary trunk feather palm.

## POGONOTIUM

Three species, from the Malay Peninsula and Sarawak.
Solitary or clustered rattans, with feathered leaves.

## POLYANDROCOCOS

Two solitary species, from Brazil, occurring close to the sea in sandy soils, in both open areas and forests.
*P. caudescens,* Buri Palm (Brazil). This palm has a sturdy trunk 4–7m (12′–22′) high with a head of about 16 feather leaves, each with around 200 pointed leaflets. These are green above and white-hairy beneath. The unbranched inflorescence has yellow flowers; as the fruits start to form it droops down below the

*Polyandrococos caudescens*
Buri Palm.

*Prestoea longipetiolata*
This genus is similar to *Euterpe*.

*Pritchardia affinis*
Lou Palm, of Hawaii

*Pritchardia affinis* var. *gracilis*
Lou-lu Palm, of Hawaii

leaves, eventually becoming a cluster of tightly bunched oval greenish-orange fruits 4–5cm (1½″–2″) long. These are fed to stock in Brazil.
**Culture and habitat** *Poyandrococos* species are not very well known, but in many ways resemble the popular *Syagrus* (Queen Palm) except for the leaflets, which are in a single plane, and the unusual cylindrical fruit bunches over 1m (3′) long which follow the rather large (for palms) yellow flowers. They grow well in both tropical and subtropical climates, in coastal regions with sand soils, but require ample water supplies. They might be welcome addition for seaside plantings.
**Sex** Monoecious. Self pollinating for seed production.
**Position indoors** No knowledge of use indoors.
**Propagation** Seeds germinate after a couple of months.

## PRESTOEA

28 solitary or clustered species, from the West Indies, Central and South America from Nicaragua south to Brazil and Peru, and growing on well-drained mountain slopes, at medium to high altitudes of 900m (3000′) or higher.
*P. longipetiolata* (Costa Rica) This species has a medium trunk with a crownshaft topped with nicely formed feather fronds, with shiny deep green leaflets. The branched inflorescence opens out of a horn-like spur; the round fruits, dark purple at maturity, are 12mm (½″) long.
**Culture and habitat** *Prestoea* species are closely allied to *Euterpe* and are found growing in similar terrain and conditions, but only at the mid altitudes of that species range. Refer to *Euterpe* for growing conditions.
**Sex** Monoecious. Self pollinating for seed production.
**Position indoors** *Prestoea* is so closely allied to *Euterpe* that the same indoor conditions may also apply.
**Propagation** Refer to *Euterpe.*

## PRITCHARDIA

37 solitary species, Pacific Islands from Fiji to Hawaii and growing on windy slopes from sea-level to 1400m (4500′) in wet forested areas; a few species grow in dry forests in positions protected from the wind.
*P. affinis,* Lou Palm (Hawaii). A fan palm with a thick, rough trunk to 8m (25′) topped with broad, heavy mid-green leaves. The stiff inflorescence of cream flowers is followed by 25mm (1″) black fruits.
*P. affinis* var. *gracilis,* Lou-lu (Hawaii). This variety has a ringed trunk to 10m (30′) high. The mid-green fan leaves 1m (3′) long, which turn yellowish as they age, have heavy white-scurfed stalks. The inflorescence hangs down and the 25mm (1″) fruits are black.
*P. eriophora* (Kauai, Hawaii). This species has large,

*Pritchardia eriophora*
A species from Kauai, Hawaii.

*Pritchardia gaudichaudii*
This species grows on Molokai Island.

*Pritchardia hillebrandii*
Loulou Lelo Palm, of Hawaii.

*Pritchardia kahukuensis*
Lo-ulu Palm, of Oahu, Hawaii.

*Pritchardia macrocarpa*
Lo–ulu Palm, of Oahu, Hawaii.

*Pritchardia maideniana*
Apparently extinct in its native habitat, this species survives at the Royal Botanic Gardens, Sydney.

well-formed bright green fan leaves with arching green stalks.

*P. gaudichaudii* (Molokai, Hawaii). This is an attractive fan palm with a thick trunk to 6m (20′) tall. The symmetrically placed leaves are green and have stalks covered in white scurf. Fruits are round and 25–40mm (1″–1½″) in diameter.

*P. hillebrandii,* Loulou Lelo Palm (Hawaii). This fan palm has a trunk to 7m (23′) high. The leaves are bluish-green with the underside and stalk white-powdery. Flowers are yellow and fruits are brownish-black, round, 15–20mm (½″–¾″) in diameter.

*P. kahukuensis,* Lo-ulu, (Oahu, Hawaii). This is a robust and handsome species having broad fan leaves with sturdy green stalks, nicely displayed.

*P. macrocarpa,* Lo-ulu (Oahu, Hawaii). This species has a ringed trunk to 10m (30′) high. The mid-green fan leaves have less thick green stalks and arch outwards.

*P. maideniana* (origin unknown). This species has a dark, rough trunk to 5m (15′) high topped by a ball of heavily segmented fan leaves. These vary in colour: the newest are bright green, older ones are yellow-green, and spent brown ones continue to hang on the plant. The inflorescence consists of clusters of orange flowers, succeeded by round brownish black fruits 12mm (½″) in diameter. This species is unknown in the wild, but two old plants are growing in the Royal Botanic Gardens, Sydney.

*P. martii,* Loulou-hiwa Palm (Hawaii). This is a compact fan palm with a 3–4m (10′–13′) trunk. The leaves are mid-green and are held stiffly. Yellow flowers

*Pritchardia martii*
Loulou–hiwa Palm, of Hawaii.

*Pritchardia pacifica*
Fiji Fan Palm.

*Pritchardia thurstonii*
Another Fijian Fan Palm.

*Pritchardia woodfordii*
In the dome at Brisbane Botanic Gardens, from Hawaii.

*Pritchardia vuylstekeana*
A species from the Society Islands.

*Pseudophoenix lediniana*
A tall species from Haiti.

*Pseudophoenix sargentii*
Buccaneer Palm.

are followed by oval greenish-brown fruits 25mm (1″) long.

*P. pacifica,* Fiji Fan Palm (Fiji, Tonga, Samoa). This impressive fan palm has a thick 10m (30′) trunk. The bright green leaves have white powdery stalks and are nicely displayed. The flowers are yellow and the round black fruits are 12mm (½″) in diameter.

*P. thurstonii* (Fiji). This is a handsome fan palm with a ringed and patterned brown trunk 8m (25′) high, topped by a crown of lovely broad, bright green fan leaves. The long flower spray arches out as much as 2m (6′) and has yellow flowers which are succeeded by small round deep red fruits 6mm (¼″) in diameter. Suits sea-coasts.

*P. vuylstekeana* (Society Islands). This is a sturdy grower with a solid trunk. The large, bright green, fan-shaped leaves have heavy stalks carrying fawn fuzz.

*P. woodfordii* (Hawaii). This interesting species is growing in the conservatory at Brisbane Botanic Gardens. It has a thick trunk topped with large deep green fan leaves with powdery white stalks.

**Culture and habitat** *Pritchardias* are medium-sized and very handsome fan palms from the tropics. Many of them are subjected to sea winds in their natural habitat, but in cooler zones than the tropics such winds may be too cool for them. They appear to adapt quite well to a warm temperate climate unless subjected to cold winds and frosts. However, it must also be remembered some species are more tender than others.

For garden plantings select a sunny, roomy position and (if away from the tropics or subtropics) a warm sheltered area with reasonably rich soil, and be sure to water regularly. If potted into large containers they are handsome tub specimens.

**Sex** Hermaphrodite. Self pollinating for seed production.

**Position indoors** No information available on indoor use.

**Propagation** Seeds germinate in 2–3 months. Protect young plants from cold and wind.

## PRITCHARDIOPSIS

One species, *P. jeanneneyi,* in New Caledonia.
Solitary trunk palm, with fan leaves.

## PSEUDOPHOENIX

Four solitary species, occurring in the West Indies and on the surrounding Caribbean coastline, growing near the coast in soft sandy soils, and inland on dry hills.

*P. lediniana* (Haiti). The handsome tapered grey trunk 12m (40′) high is nicely ringed, and crowned with a rather open head of feather leaves to 3m (12′) long with very fine leaflets. The crown and the bases of the leaf stalks are silvery grey.

*P. sargentii* Buccaneer Palm (Florida and Central America). This species has a grey, ringed, bulging trunk to about 8m (25′) topped with a wide crown of stiff, blue-green feather leaves. The 45cm (18″) inflorescence hangs from the lower leaves and bears a broad cluster of two-seeded, kidney-shaped red fruits 30mm (1¼″) wide by 18mm (¾″) long, each containing two or three seeds. Tolerates salt water and salty coastal winds.

*P. sargentii* subsp. *saonae* (Hispaniola). This form has only minimal differences from the above subspecies.

*P. sargentii* ssp. *saonae* var. *navassana* (Navassa Island). This variety is much taller-growing than the above one, and its trunk is also less bulging. It also appears from the distance between the lower rings that its early growth is much more rapid.

*P. vinifera,* Cherry Palm (Hispaniola). This has a broader, more tapered trunk than the other species and grows to about 10m (30′). It carries about 10 leaves around 3m (10′) in length, most attractive with their silvery stalks. The inflorescence is longer, to 90cm (3′), and flowers are followed by red fruits 30mm (1¼″) long.

**Culture and habitat** *Pseudophoenix* species are native to the Caribbean area of America and the West Indies. They are often described as resembling small royal palms, but their trunk rings soon identify them. One species, *P. vinifera* (Cherry Palm) from Haiti, may be endangered on that island because the local people extract its sweet juice for wine making, but at this stage it is fortunately still quite common.

*Pseudophoenix* species are not widely grown away from their natural home, unlike their taller look-alikes, the royal palms. However, with the range of

*Pseudophoenix sargentii* subsp. *saonae* var. *navassana*
This variety comes from Navassa Island.

*Pseudophoenix sargentii* subsp. *saonae*
The trunk is sturdier in this subspecies.

*Pseudophoenix vinifera*
Cherry Palm is used for making wine.

*Ptychococcus elatus*
A slender palm with drooping leaflets.

*Ptychococcus lepidotus*
Tall specimens in the foreground. The rear palm is *Arenga undulatifolia.*

plantings in some of our botanic gardens being broadened, this may soon be rectified. They will be a handsome addition.
*P. sargentii* comes from subtropical Florida and the Caribbean, and there should be no problems in growing this species in these zones, where it should be useful for coastal plantings. Perhaps one day it will be seen more plentifully than the now common royal palms. The unusual tapering trunks are very eye-catching and a talking point.
*Pseudophoenix* will grow in all the warmer zones, as well as in frost-free warm temperate areas, but here cold coastal winds may be detrimental. They prefer an alkaline or even saline soil with good drainage and will even tolerate low and erratic rainfall once established.
**Sex** Polygamous or hermaphrodite, carrying either separate male and female or bisexual flowers. Self pollinating for seed production.
**Position indoors** No information on whether they are used indoors.
**Propagation** Seeds take from 1 to 2 months to germinate, but growth is very slow and this is probably the reason why they are not more widely grown for the nursery trade. Plants are hardy. The seeds are long-lived, keeping for up to 2 years, and in their natural habitat will float away to distant shores.

## PTYCHOCOCCUS

Seven solitary species, from New Guinea and the Solomon Islands, growing in lowland rainforests and on mountain ridges. The wood is tough and is used for bows and arrows as well as for building.
*P. elatus* at Fairchild Gardens, Florida. A slender palm to 4–5m (12′–15′) on a slim trunk, the feather leaves have leaflets with cut-off ends and droop from the mid rib.
*P. lepidotus* (New Guinea). This slender palm to 5m (15′) has a thin grey trunk topped with an open head of feathery fronds. The leaflets have blunt ends. There is a crownshaft covered in fine hairs that gives it a grey appearance. The branching inflorescence has green flowers followed by oval red fruits 4–5cm (1½″–2″) long.
*P. paradoxus* (Irian Jaya). This is a small graceful palm with a slender trunk only growing to about 4m (12′) high. The trunk is topped by a crownshaft and a small head of fine feathery leaves, creating an effect of lightness. The red fruits resemble a large olive in shape. This palm was previously known as *Ptychosperma paradoxus.*
*P.* sp. *7L533,* at Foster Gardens, Hawaii (New Guinea and Solomons). This palm appears to be *P. lepidotus.*

**Culture and habitat** *Ptychococcus* is very closely allied to *Ptychosperma* and the same cultural conditions would apply to both. However, as *Ptychococcus* originate mostly from hotter regions, care should be taken not to expose them to low temperatures. They grow well in Hawaii, Brisbane and Sydney. They need to be well watered and sheltered from cold, especially when being grown in temperate areas. Here they

may be happier grown amongst other palms and plants, in a rainforest setting as an added protection against cold winds.
**Sex** Monoecious. Self pollinating for seed production.
**Position indoors** Nothing is specifically known, but it is thought that they would be similar to *Ptychosperma* indoors.
**Propagation** Seeds are said to be easy to germinate.

## PTYCHOSPERMA

28 solitary or clustered species, centred in New Guinea, with a few in surrounding areas of the Moluccas, northern Australia and nearby Pacific Islands, growing in a variety of habitats, from swamps to mountains. They have long, slender trunks, either single or clustered, depending on the species. Their bunches of bright red fruits are borne plentifully and add a touch of colour.
*P. ambiguum* (Irian Jaya). This tropical clumping palm has slender trunks to 5m (15′) under ideal conditions, topped with a head of short 1.3m (4′) leaves. The compact clump is quite leafy at the base, making the species ideal for pots in warm areas. The black fruits make a striking contrast with the red stems.
*P. burretianum* (New Guinea). Grows on clustered trunks. The feather leaves have broad caryotoid leaflets, almost bat-wing in shape.
*P. elegans,* Solitaire Palm (north-eastern Queensland). This feather palm has a slender grey ringed trunk to about 10m (30′) high, with a short crownshaft and bright green, cut off leaves paler beneath. The white flowers are followed by a large bunch of oblong bright red fruits 15–20mm (¾″) long. Common around Cairns.
*P. ledermannianum* (syn. *Ponapea ledermannianum)* (Ponape and Kusaie Islands). This species has a solitary ringed trunk to 10m (30′) high and much resembles *P. elegans.* The leaves are 2–3m (6′–10′) in length with a greyish leaf base. The fruiting stem is conspicuously branched and yellow; the red fruits are oblong, 4–5cm (1½″–2″0) long.
*P. macarthurii,* Macarthur Palm, Hurricane Palm (Cape York Peninsula to New Guinea). This is a feather palm with clumping, slender, grey ringed trunks 3–6m (10′–20′) high, topped with a crownshaft. The cut-off leaves are dull green and paler beneath. The cream flowers are succeeded by a bunch of bright red fruits 12mm (½″) long. *P. macarthurii* has a few variants including a variety with finer leaves and another named *P. mac.'Robustifolia'.*
*P. macrospadix** (New Guinea). This is a tall clustering palm similar to *P. macarthurii,* but also resembling *P. microcarpum* in its toothed leaf ends; it grows taller than the latter species.
*P. microcarpum* (New Guinea). This compact species growing to 5m (15′) has clustered trunks topped with bright green feather leaves with clustered leaflets. Red flowers are followed by black fruits.
*P. pullenii* (New Guinea). This is a small grower with a ringed trunk. The leaf stalks are grey at the base and the leaflets are triangular.

* *P. macrospadix* is now a synonym of *P. microcarpum*

*Ptychococcus paradoxus* (juvenile)
A slender palm used for bows and arrows.

*Ptychosperma ambiguum*
A multitrunked species from Irian Jaya.

*Ptychosperma burretianum*
A slender cluster palm with caryotoid leaflets.

*Ptychosperma elegans*
Solitaire Palm.

*Ptychosperma ledermannianum*
This palm was previously classified as a species of *Ponapea.*

*Ptychosperma macrospadix*
This species resembles *P. macarthurii,* but has narrower leaflets.

***P. salomonense*** **(Solomon Islands). This is a single-trunked 12m (40′) variety, resembling *P. elegans,* but with a head of longer and broader leaves on shorter stalks. The young leaf sheaths have an early covering of white scurf. The spray of bright red fruits is decorative.**

***P. sanderianum*** **(New Guinea). This species is a clumping palm very similar to *P. macarthurii* but with narrower and shorter leaflets and smaller fruits. It makes an attractive clustering potted palm.**

***P. waitianum*** **(New Guinea). This species, with a single trunk under 5m (15′) high, is shorter than others. The crownshaft is topped by a graceful crown of feather leaves that are shorter but have broader segments than most varieties; leaves may be red when juvenile. The red flowers are followed by black fruits 2cm (3/4″) long. Likes shade and protection from cold.**

**Culture and habitat** In nature *Ptychosperma* species occur in rainforest and mountain valleys or the fringes of swamps.

They are easy to grow in the tropics and quite hardy, resisting strong coastal winds. Even in the subtropics they will grow quite well away from cold and cold winds, but may get a few leaf brown spots during winter. These do not appear on glasshouse-grown specimens used for indoor growing. Out of doors in the subtropics it would be best to select a suntrap area. In warm temperate areas, the extreme fringe of their outdoor range, they would need a suntrap site protected from cold winds in a coastal frost-free area; select a warm position under the protection of trees or taller palms. Water freely and often. Protect from strong, drying winds.

*Ptychosperma macarthurii*
Macarthur Palm or Hurricane Palm.

*Ptychosperma macarthurii*
These fine specimens are growing in Singapore.

*Ptychosperma macarthurii* (right), *P. 'Robustifolia'* (left)
Shows the difference at juvenile age.

*Ptychosperma microcarpum*
**The leaflets are strongly clustered.**

*Ptychosperma pullenii*
**A small grower from New Guinea.**

*Ptychosperma salomonense*
**A Solomon Islands species.**

*Ptychosperma sanderianum*
**The leaflets are smaller than in *P. macarthurii*.**

Clustering varieties are planted singly, but single-trunked varieties are best planted two or three in one pot to form a clump. For blemish-free foliage it is best to grow them in a glasshouse or in a protected warm position. Keep well watered and feed occasionally with a complete plant food to maintain lush growth. Repot as required, using an indoor potting mix. 200–400mm (8″–16″) pots are best sizes.
**Sex** Monoecious. Self pollinating for seed production.
**Position indoors** Select a good naturally lit position for longer indoor life. Keep moist, and take outdoors regularly for a hose-down, avoiding direct sun. 200–300mm (8″–12″) pots are best indoor sizes.
*Ptychospermas* are now grown in Australia by commercial nurseries as a decorative indoor palm. For this purpose they appear as hardy as *Howea (Kentia)*, apart from the problem of leaf blemishes in cooler climates. Their lush green colouring is attractive indoors, but to some, the cut-off appearance of the leaf ends is unattractive, and a sales deterrent in retail nurseries.
**Propagation** Fresh seeds germinate in 2–3 months. Clustering types may also be divided.

*Oncosperma* sp.

## RAPHIA

28 solitary or clustered species, from tropical Africa, northern Madagascar and equatorial South America growing in humid swamplands.

*Ptychosperma waitianum*
A neat, compact grower with reddish juvenile leaves.

*Ravenea hildebrandtii*
A taller slimmer grower.

*Ravenea moorei*
A handsome palm.

*R. vinifera,* Raffia Palm, African Bamboo Palm (Nigeria). This palm has multiple trunks which are usually short and stout, but the tall feather leaves may end up reaching as high as 10m (30′). The inflorescence is unusual, having the appearance of a sausage up to 3m (10′) long. Both male and female flowers are present, and oval brown scaly fruits 7cm (2½″) long follow. The local people make a wine by cutting off the inflorescences at an early stage and collecting the juice. After seeding, the stem dies.

**Culture and habitat** Raffia is collected from the leaves of these palms and used for a multitude of purposes. Although today raffia is partly replaced by manufactured fibres. I can well remember, as a lad in the family nursery, wrapping seedlings in newspaper and tying these with raffia. Now the seedlings are in punnets, and very few younger folk would know what raffia was.

*Raphia* palms are very large-growing, perhaps too large for the average garden. They require a tropical climate, but could survive in warm sheltered positions from the humid tropics to the subtropics (where they would need to be in a frost-free, sheltered spot). Most of them grow in swamplands where humidity is constant, so in any zone high humidity is essential. This may be aided by regular watering.

**Sex** Hapaxanthic (each stem is monoecious and dies after fruiting). Self pollinating for seed production.

**Position indoors** No information available on indoor use.

**Propagation** Germination is slow and patchy, taking 3–6 months and requiring plenty of water.

*Ravenea glauca*
A low dainty palm from Madagascar.

*Ravenea robustior*
A Feather Duster Palm.

# RAVENEA

10 solitary species, eight from Madagascar and two from the Comores Islands, and growing in tropical rainforests from sea-level to 2000m (6000′).

*R. glauca* (Madagascar). This small, very attractive palm has a solitary trunk 3–6m (10′–20′) high. The leaves have closely spaced narrow leaflets.

*R. hildebrandtii* (Comores Islands). This is a taller-growing species 10–12m (30′–40′) with a ringed trunk. The leaf bases are brown-fuzzy and the leaf stalks are pale. The feather leaves are a pleasing shape.

*R. moorei* (Grand Comore Island). A taller, sturdy species with a solid ringed trunk. The long, graceful leaves have greyish stalks covered in rusty brown fuzz that contrasts with the beautifully displayed narrow green leaflets.

*R. robustior* (Madagascar). Best described as a sturdy 'feather duster' palm, this species has a robust trunk growing to a height of 15–25m (50′–80′), depending on the surrounding vegetation, topped with an upright, arching head of 12 to 15 feather leaves 3–4m (10′–12′) long. The flower sprays are covered in a silvery down and the fruits are small 16mm (⅝″), and red when ripe. This palm grows in wet forests, rising above the other vegetation.

**Culture and habitat** *Raveneas* appear to be sturdy and robust growers although they have hardly been tried in cultivation. Those who have seen *R. robustior* in Madagascar describe it as a noble palm in the wet forest areas where it grows, standing above the competing vegetation like a giant feather duster.

While it is untried commercially in nurseries, some specimens have been grown in the USA, in Florida and California. The climates of these places are subtropical and warm temperate, respectively. The same success should be possible in similar regions. *Raveneas* would require a moist position, or a constant supply of moisture, perhaps similar conditions to those needed by *Roystoneas* (Royal Palms). In cooler regions frost-free suntraps would be a must, plus special care to keep them out of cold harsh winds.

**Sex** Dioecious. Need a male and a female for seed production.

**Position indoors** No information available.

**Propagation** No information on seed raising is available, but it is expected that bottom heat would be required for best results, along with the humidity of a plastic or glass 'tent' covering.

# REINHARDTIA

Six solitary or clustered species occurring from Mexico to north-western Colombia, and one from the Dominican Republic. They are understorey palms from tropical rainforests, but two are found at altitudes of 1000–1500m (3000′–5000′).

*R. gracilis,* Window Palm (southern Mexico, Central America). This species has four varieties all differing in height and leaf formation and all have window splits. *R. gracilis* leaves bear from 8–11 leaf veins each

*Reinhardtia gracilis*
Has four varieties differing in form.
Window Pane Palm.

*Reinhardtia gracilis* var. *gracilior*
Little Window Palm.
An excellent pot or fernery palm.

*Reinhardtia simplex*
A colourful small species.

*Rhapidophyllum hystrix.*
Needle Palm comes from south-eastern USA.

side of the leafstem. It has taller growth, up to about 1.3m (4′).

*R. gracilis* var. *gracilior,* Little Window Palm (southern Mexico to Central America). This species attracts the most interest, with its slender trunks to about 1m (3′), covered in brown fibre and old leaf bases. The small, heavily (from 14–22) veined deep green leaves have 'chewed off' ends and small holes at their bases. Inflorescences come amongst the leaves and are followed by small 15mm (½″–⅝″) black fruits on red stems.

*R. simplex* (Honduras to Panama). This is an unusual small palm under 1m (3′) tall, with multiple thin trunks. The leaves have short stalks bearing three small shiny green simple leaflets. The 30cm (12″) long inflorescences are red at the end and later bear small green to black fruits).

**Culture and habitat** In nature, *Reinhardtias* inhabit the floor of tropical rainforests, away from direct sunlight and under the cover of other trees. They may be grown outdoors in the tropical, subtropical and warm temperate zones. However, beyond this they would require the protection of a greenhouse. Because of their small size, pot culture may be preferred.

**In the garden** Select a warm position to match the natural environment. Make sure the soil is well drained and dig in some humus or indoor potting mix. Keep very well watered, regularly hosing-down the foliage.

**In pots** Pot on as required, in indoor potting mix, to 200–300mm (8″–12″) pots, only larger if the plant requires it. Grow in a glasshouse or some other warm; shaded, favourable position. Feed occasionally with a complete plant food to maintain good green colour.

**Sex** Monoecious. Self pollinating for seed production.

**Position indoors** *Reinhardtias* are rather fussy indoors, requiring good light, warmth with humidity, and regular observation in order to detect when they need resting out of doors. Keep well watered, but not standing in a saucer of water. Do not polish with white oil as this burns the leaves. I have killed one by using a white oil aerosol spray.

**Propagation** Seed germinates within 6 weeks with bottom heat. Plants may also be divided carefully, with roots on each cluster division.

## RETISPATHA

One species, *R. dumetosa,* from Borneo.
Clustered spiny rattan, with feathered leaves.

## RHAPIDOPHYLLUM

One clustering species from south-eastern USA, growing in wet swampy areas, or in sunnier (but still moist) situations in limestone areas.

*R. hystrix,* Needle Palm (south-eastern USA). This palm is also known as the 'vegetable porcupine' because of the appearance of its trunk. It is a densely

*Rhapis excelsa 'Tenzan'*
The leaves have about four segments.

*Rhapis excelsa*
Lady Palm does best in shade, excellent in tubs.

*Rhapis* cv. *'Kinsho'*
Variegated forms can be propagated only by division.

*Rhapis* cv. *'Akatsuki'*
This form is rare and can be multiplied only by division.

*Rhapis* 'Burmese Hybrids'
An assortment from commercially available seeds.

*Rhapis* sp. (at Fairchild Gardens, Florida, USA)
Distinctly eye-catching, and thought to be *R. laoensis*

suckering palm with many trunks 1m (3′) or more high, massed with long spines and fibres. The green fan leaves have narrow green stalks 1–1.5m (3′–5′) long. The inflorescence is held within the leaves and the globose brown furry fruits are 2cm (¾″) long.
**Culture and habitat** Because *Rhapidophyllum* grows so slowly from seed, it is rarely seen in nurseries. In nature it forms thick clumps and in past years trunks were cut as curiosities, endangering the species. This practice has now stopped. This is a hardy pal that may be grown in the warm temperate zone and all warmer zones. The trunks may rot at the base so that they lean or even lie on the ground. A warmer, sunny situation will promote more rapid growth. Regular watering and alkaline soils also help.
**Sex** Flowers may be either bisexual or unisexual. Self pollinating for seed production.
**Position indoors** No knowledge on indoor use.
**Propagation** Seeds are slow, taking up to 6 months to germinate. Division of trunks or suckers may be faster.

## RHAPIS

12 clustering species, occurring from southern China to Thailand, and growing as understorey palms in dry evergreen forests and on limestone hillsides.
*R. excelsa,* Lady Palm (southern China). A densely clumping palm with slender trunks to 5m (15′) tall covered with brown fibres. Segmented fan leaves

*Rhapis minor*
A smaller species with slim trunks.

persist down the trunk. Short spikes of cream flowers appear among the upper leaves, followed by one- to three-seeded white-waxy 8mm (⅜″) fruits, if pollination occurs; single-seeded fruits are round.

*R. excelsa* 'Tenzan'. This cultivar has fewer leaf segments (four or five) and less fibre on the stems.

*R. cv 'Akatsuki'.* A rare and attractive cultivar, with ivory to cream stripes.

*R. cv 'Kinsho'.* This cultivar resembles *R. humilis* in leaf shape and form, but has variegated leaves. Reproduced by division only.

*R. humilis,* Slender Lady Palm (China). Similar to *R. excelsa,* but more slender and compact, forming a thicker clump, and with more closely woven fibres on the stem. The leaf segments are more numerous, more pointed and more deeply cut, and droop lower. Flower stalks are longer. Female flowers are unknown on this species hence plants are propagated only by division.

*R. minor* (A gift to Fairchild Gardens from Satake of Japan. Not fully evaluated yet). In this species the multiple stems lack the fibres of the other species. In addition, the leaves have fewer segments, averaging three.

*R. subtilis* (Thailand). A fine clump of this less well-known species can be seen growing under a canopy of trees at Brisbane Botanic Gardens.

*R.* sp. (at Fairchild Gardens and thought to be *R. laoensis.* Awaiting proper identification). This species differs from both *R. excelsa* and *R. humilis* in the narrow leaf segments and the taller size of the trunks.

*R. 'Burmese Hybrids'.* Seeds of these are offered to the commercial trade and, when grown, produce a mixture of types, including forms resembling *R. excelsa* and *R. humilis,* and other variations.

**Culture and habitat** *Rhapis* or Lady Palms are highly prized as potted palms, and large specimens are almost status symbols, like Rolls Royce or Mercedes cars. The two widely grown species are somewhat confused these days by the introduction of hybrid seeds. These are often called 'Burmese Hybrids' etc. They result from available seed crosses.

*Rhapis* are adaptable to both outdoor and indoor growing. In nature they are found in higher-rainfall forests in tropical to subtropical areas.

**In the garden** Lady Palms will tolerate some sun, but yellow if the sun is too strong. In more shaded areas they remain a rich green. Keep well watered. Dig in some humus or potting mix before planting. A shaded warm position is best and high humidity should be maintained by hosing regularly. They will adapt to the tropics, and will also grow well in the temperate zones if protected from frosts.

**In pots** Lady Palms are suitable for indoor and outdoor pot growing. Clumps may be divided when large enough, and repotted again. The divisions should initially be planted in pots that are not too large, then progressively potted on as required. Suitable display sizes are between 200mm (8″) and 500mm (20″). A large clump is a treasure. Use an indoor potting mix, and feed occasionally with a complete plant food. Outdoors, verandas or protected areas are best.

**Sex** Dioecious (single sexed) or polygamous (male

*Rhapis humilis*
Slender Lady Palm forms a thick, slender clump.

*Rhapis subtilis*
This Lady Palm from Thailand lacks the woven fibres on the stem.

*Rhopalostylis baueri*
Norfolk Palm.

*Rhopalostylis sapida*
Feather Duster Palm or Nikau Palm.

and female flowers mixed with bisexual). Rather erratic in seed production.

**Position indoors** A sunroom or a brightly lit window area would be best. Keep moist, but not in overly wet saucers. Take outside regularly and hose-down, but do not leave outside too long, or they may become sunburnt. If light is too poor, tip drying will occur. Feed occasionally with complete plant food.

**Propagation** Seeds germinate in 2–3 months depending on the temperature. Commercial seed supplies may produce variable offspring. Propagation by division of clumps is also much practised. It has the advantage of giving plants of good height, as growth is very slow from seeds. The variegated forms are very expensive and scarce as they can only be multiplied by division, which is very slow; hence production is limited.

## RHOPALOBLASTE

Six species, from the Nicobar Islands, Malay Peninsula, Moluccas, New Guinea and Solomon Islands.

Solitary or clustered trunks, and feathered leaves.

## RHOPALOSTYLIS

Three solitary species from New Zealand and Chatham, Norfolk and Kermadec Islands, usually growing near the sea in mild lowland forests receiving good rainfall.

*R. baueri*, Norfolk Palm (Norfolk Island, Kermadec Islands). This handsome feather duster palm has a closely ringed, grey trunk to 10m (30″) topped with a short plump crownshaft. The attractive head of feather leaves is more spreading than that of *R. sapida*. A branched flower inflorescence, from below the crownshaft, is followed by oval red fruits 18 mm (¾″) long.

*R. sapida*, Feather Duster Palm, Nikau Palm (New Zealand, Chatham Islands). This feather palm has a grey ringed trunk to 6–8mm (20″–25″), topped by a short, bulbous crownshaft from which the almost stalkless leaves grow. The leaves are quite broad, deep green with closely set narrow leaflets pointing upwards, giving the impression of a large feather duster. Short inflorescences appear from below the crownshaft; the flowers are purple to mauve and the bright red fruits are round and 12mm (½″) in diameter.

**Culture and habitat** These New Zealand palms are the most southern-growing in the world. They are hardy in all but extreme climates and will tolerate some frost in the cool temperate zone. They may also be grown in the warmer zones excluding the humid equatorial tropics and drier hot inland areas. In hotter regions they would prefer a cooler elevation or a location fanned by coastal breezes. The genus is closely related to *Hedyscepe*.

**In the garden** *Rhopalostylis* species are quite hardy, though they suffer from leaf burn if exposed to hot drying winds. Coming from dense forest areas not

far from the sea, they prefer sunny to semishaded positions in mild climates with warm temperatures and with an ample water supply. Dig in some compost or potting mix over 1–2 square metres (11–22 square feet) before planting out.
**Sex** Monoecious. Self pollinating in seed production.
**Position indoors** I find these an excellent indoor palm, both hardy and attractive. Sunrooms and semi-open areas are best. If using fully indoors, select a naturally bright position, keep well watered, and perhaps take the plant outdoors weekly for a hosing-down.
**Propagation** Fresh seeds germinate in 2–3 months.

## ROSCHERIA

One solitary species, from The Seychelles, growing undercover as a second-storey species in forests and on mountainous hillsides from 500–750m (1500′–2500′).
*R. melanochaetes,* Latanier Hauban. The thin trunk of this palm is armed with rings of black spines when young, but almost bare at maturity. There is a brownish crownshaft armed with black spines and a head of eight to 10 irregularly feathery green leaves 1–2m (3′–6′) long on armed stalks. The branched inflorescence 0.6–1.5m (2′–5′) long bears small round red fruits 6–7mm (¼″) long at maturity.
**Culture and habitat** This is a rare palm, little grown or known around the world, hence information is

*Roscheria melanochaetes*
Latanier Hauban, has a spiky trunk when young.

*Roystonea elata*
Florida Royal Palm. Taller than *R. regia,* this species grows to 30m (100′).

*Roystonea oleracea*
Caribbean Royal Palm. Tapering trunk to 40m (120′), the leaflets are held flat.

*Roystonea princeps*
Jamaican Royal Palm.

*Roystonea regia*
Cuban Royal Palms, fronting Brisbane City Hall.

sparse. It is a medium-sized grower remaining below tree height in its native habitat. It develops aerial roots and was at first thought to be a type of *Verschaffeltia,* also from mountainous regions in The Seychelles. The soils it grows in are rich and spongy with organic matter; moisture is high and temperatures are tropical. Specimens are growing well in the Rare Plants House at Fairchild Tropical Gardens in Florida. I daresay it would also be worth a try wherever *Vershaffeltias* have been successfully established.

*Roscheria* should be a little less cold-sensitive than *Verschaffeltia* because of the higher elevation of its natural habitat, as well as being less dependant on high humidity. There is little authenticated information about its growth outside the tropics, but it would probably also grow in warmer subtropical areas under cover. If you are trying it in the warm temperate zone, it should also be given protection from frost and cold winds, perhaps in a glasshouse or similar situation. Other requirements are rich soil, absence of cold winds, a surface mulch and ample regular cover-waterings.

**Sex** Monoecious. Self pollinating in seed production.
**Position indoors** No information is available.
**Propagation** There is no precise information, but it is thought that it would require similar heat and humidity to *Verschaffeltia.*

# ROYSTONEA

10 to 12 solitary species from the Caribbean Islands and the surrounding mainland and growing in the rich lowland soils. Most of the original habitat is now cleared for agriculture, but the palms are widely grown throughout the world.

*R. altissima,* Mountain Cabbage Palm (Jamaica). This Royal Palm is found at altitudes on the island of Jamaica. It is similar to other species.

*R. boringueana,* Puerto Rican Royal Palm (Puerto Rico). This beautiful palm is distinguished from the other royal palms by the shiny upper surface of the leaves, plus the denseness of the flowers in the inflorescence. The oblong fruits are yellow-brown.

*R. elata,* Florida Royal Palm (Florida). This species from the swampy everglades was once confused with *R. regia.* It grows taller, to 30m (100′), with a ringed ash-grey trunk, a long green crownshaft and a nice crown of dark green feather leaves. The inflorescence is longer and looser and the fruits are rounder and purplish.

*R. oleracea,* Caribbean Royal Palm (southern Caribbean area). This, the tallest of the species, has a tapered greyish trunk to 40 m (120′) high topped by a long green crownshaft and attractive head of leaves, held above the crownshaft. The cream flowers appear below the crownshaft, followed by 15–20mm (¾″), oblong purplish black fruits.

*R. princeps* (Jamaica), "Jamaican Royal Palm". Grows on a more slender trunk, tapering upwards. The head of leaves is sparser than most and the narrow leaflets droop. The inflorescence is very branched and broom-like on yellow stems.

*R. regia,* Cuban Royal Palm (Cuba). This is the most grown and the most outstanding species. It has a smooth greyish-white tapering trunk to 25m (80′) high, swollen at the top and base, and topped by a long green crownshaft. The head of plume-like leaves hangs low enough to cover the crownshaft. The inflorescences are much branched and produce almost round 8–12mm (⅜″–½″) purplish black fruits.
*R. venezuelana,* Venezuelan Royal Palm (Venezuela). 'Magnificent' is most people's first thought at seeing this large palm. The trunk is heavy and tall, about 25–30m (80′–100′) high. The strong head of leaves is displayed above the crownshaft. The flower and seed sprays are compact and thick and very tidy.

**Culture and habitat** The magnificent royal palms with their stately trunks and feathery tops are majestic even amongst palms. These stately specimens are widely used for lining roadways or walks where the smooth grey-white trunks are seen to their best advantage. Nevertheless, a single specimen is also a garden or park gem.

In nature they grow in rich lowland forest areas of high rainfall or even in swampy areas close to the sea. While coming from the tropics, they have adapted to other climates well. It is said that the more water they get, the quicker they grow. There are a number of varieties, all from the Caribbean Islands and nearby parts of America. The most grown is *R. regia,* the Cuban Royal Palm, popular for its oddly shaped trunk and regal stature.

**In the garden** Royal palms require the warmth of

*Roystonea borinqueana*
The trunk bulges above the middle.
Puerto Rican Royal Palm.

*Roystonea venezuelana*
Venezuelan Royal Palm is a sturdy palm with a thick trunk.

*Roystonea altissima*
Mountain Cabbage Palm of Jamaica.

*Sabal causiarum*
**Puerto Rican Hat Palm.**

*Sabal domingensis*
**Palmetto Palm, from Hispaniola**

*Sabal bahamensis*
**A large, sturdy grower, from the Bahamas.**

*Sabal mauritiiformis*
**A species from Central America.**

the tropics and subtropics. Given warm protected pockets and sheltered frost-free positions they can be grown quite successfully in both the warm and cool subtropics, if placed in a position sheltered by the house or taller trees from cold or drying winds. Incorporate some compost or potting mix into the soil before planting. Keep well watered to promote more rapid growth, and fertilise occasionally with a complete plant food. They grow in rich soils in nature.
**Planted in pots** Royal palms are tall growers, and not very suitable as permanent potted palms. If you want potted specimens the pots would have to be very large, in excess of 1m (3′) and the plants would need regular daily watering.
**Sex** Monoecious. Self pollinating in seed production.
**Position indoors** Unsuitable as indoor palms, being too tall and spindly as juveniles.
**Propagation** Seed germinates within 2 months, best with bottom heat.

## SABAL

14 solitary species, occurring from southern USA to Venezuela, in dry open lands, sandy coastal regions or swamplands.
*S. bahamensis* (Bahamas) This species has large, stiff blue-green fan leaves with stalks almost 2m (6′) in length.
*S. bermudana* (Bermuda). This palm has a thick patt-

*Sabal minor*
Dwarf Palmetto Palm, from south-eastern USA.

*Sabal bermudana*
A hardy and sturdy fan palm, from Bermuda

*Sabal princeps*
A very large grower, from the West Indies.

*Sabal mexicana*
Mexican Palmetto or Texas Palmetto.

erned trunk topped with a compact head of blue-green fan leaves. It is quite hardy in climates from tropical (Hawaii) to warm temperate with light frosts (Melbourne). It requires good drainage, and is sun-tolerant early.

*S. causiarum*, Puerto Rican Hat Palm (Puerto Rico, Virgin Islands). This is a giant, with a tall trunk often 60 cm (2′) thick and 15m (50′) high. The fine specimen growing in Brisbane Botanic Gardens is much taller. About 40 drooping fan leaves 1.3–1.6m (4′–5′) long are bright green, or sometimes dusty blue-green. The sprays of fragrant white flowers, over 2m (6′) long, are followed by small 7–12mm (¼″–½″) blackish-brown fruits.

*S. domingensis*, Palmetto Palm (Hispaniola). This is a large fan palm with a heavy, thick trunk to 20m (60′) high. There is a full, round head of dark blue-green fan leaves. The clustered inflorescence is hidden amongst the leaves and later bears small black fruits. This species will stand up to hot, dry inland climates.

*S. mauritiiformis* (Colombia, Venezuela Trinidad). This tall grower has a slender 10m (30′) trunk and drooping mid-green fan leaves. The inflorescences extend beyond the leaves.

*S. mexicana* (syn. *S. texana*), Mexican Palmetto, Texas Palmetto (Texas, Mexico). This species has a thick trunk to 60cm (2′) wide and 20m (60′) high. The 15 to 30 light green leaves have blades 2m (6′) long very stiffly displayed on slightly shorter stalks. Inflorescences are about equal to the leaves; the fragrant

*Sabal uresana*
Sonoran Palmetto, from Mexico.

*Sabal palmetto*
Palmetto Palm, from south-eastern USA.

flowers are white, followed by round 12 mm (½″) brown fruits; some of which may be double. This species from the Rio Grande region is similar to *S. palmetto.*

*S. minor* (syn. *S. adansonii*), Dwarf Palmetto, Blue Palmetto (south-eastern USA). The short trunk, rarely exceeding 2m (6′) and rough with leaf bases, is topped with 8 to 20 blue-green fan leaves. The older spent leaves collapse and droop downwards. The inflorescences extend above the leaves; flowers are white, and fruits black and 7mm (¼″) long.

*S. palmetto,* Palmetto, Cabbage Palm (south-eastern USA). The trunk of this species, patterned with old leaf bases, is 6–25m (20′–80′) tall and about 30–45cm (12″–18″) in diameter. It is crowned with 30 large, green to blue-green fan leaves with a blade 2m (6′) long on a long stalk. An inflorescence the same length is followed by roundish fruits, about 10 mm (⅜″) in diameter and shining brown to black.

*S. princeps* (West Indies). This is the largest species, with a thick trunk to 20m (60′) and a massive head of 2m (6′) wide blue-green fan leaves. Inflorescences are borne amongst the leaves; white flowers are followed by 15mm (⅝″) brownish-black fruits.

*S. rosei* (Mexico). This palm has a sturdy trunk covered with old leaf bases, topped with a loose head of drooping silver-green fan leaves. The inflorescences hang below the leaves.

*S. uresana,* Sonoran Palmetto (Mexico). This species differs in that the bluish green leaves are almost fully segmented, making it particularly attractive as a juvenile plant. It later becomes less blue and grows a trunk.

*S. yapa* (Cuba). This handsome species has blue-green fan leaves that are silver-blue beneath. The inflorescences extend well beyond the leaves.

**Culture and habitat** *Sabal* species, the Palmetto Palms, are hardy, sun-loving fan palms. The different species originating from tropical and subtropical regions adapt to sandy soils as well as to swampy conditions, and will tolerate light frosts in the cooler zones. They grow on broad trunks, sometimes almost a metre (3′) across, with the old leaf bases remaining attached, in most cases, over most of the height of the trunk. The brown spent leaves also persist for some time.

*Sabals* are very hardy and much like the *Livistonas* in habit and spread, so they require room to be seen to best advantage. Most grow tall heavy trunks. They are best in good soil and should be kept well watered, but will adapt to drier regions as well.

*Sabal* species are hardy in pots. *S. minor,* being the smallest, is best for pot growing, as the others soon outgrow containers. Pot on as necessary, one size larger at a time, using a good potting mix. Best display sizes are 300–500mm (12″–20″) pots, or even larger for large specimen plants. Keep evenly moist and feed occasionally with a complete plant food.

**Sex** Hermaphrodite. Self pollinating in seed production.

**Position indoors** *Sabals* are rarely used indoors, but they are quite suitable for this purpose, if given good light, even a sunny corner. The best species to use would be *S. minor.* Potted specimens should be kept

*Sabal rosei*
The trunk is patterned with old leaf bases.

*Sabal yapa*
The blue colouring of this species is attractive. From Cuba.

*Salacca zalacca*
Salak has edible fruits.

*Satakentia liukiuensis*
Satake Palm is most attractive.

evenly moist and fertilised occasionally for attractive growth. Remove to outdoors at first signs of distress.
**Propagation** Seeds germinate within two months and seedlings should be potted before they become too deep-rooted.

## SALACCA

15 clustered species, occurring from Burma to India to China, then southwards to Indonesia and the Philippines, growing as understorey cluster palms in tropical rainforests.
*S. zalacca* (syn. *S. edulis*), Salak (Indonesia). This palm forms a large and impenetrable cluster with multiple underground trunks which send up 4m (12′) high feather leaves with very thorny stalks. The leaves stand erect and are green on one side and whitish on the other. The inflorescences are held erect amongst the leaves and followed by much-prized pear-shaped 7cm (2¾″) fruits with brown spiky scales and yellow edible flesh; these usually have three seeds.
**Culture and habitat** These palms grow in rich soils, sometimes in swampy areas or along streams, and form impenetrable barriers. They are rarely grown as garden specimens, although they can be imposing and interesting clumps if grown away from thoroughfares. Although they are tropical, specimens also grow in the subtropics, doing well in Brisbane Botanic Gardens. They will not tolerate cold and should be sheltered from cold winds. Humidity from a surface mulch is beneficial, along with regular watering.

Although rather forbidding as garden specimens, *Salacca* species are quite interesting as pot-plants; not that nurseries are offering them for sale. This way they can be kept under control. Their tolerance of sun from an early age makes them useful.
**Sex** Dioecious. Need a male and a female for seed production.
**Position indoors** No information is available.
**Propagation** It is reported that seeds may be difficult to germinate, but they have been raised in 1 month; perhaps bottom heat is required.

## SATAKENTIA

One solitary species, from the Ryukyu Islands and growing on the hill slopes. Previously known as *Gulubia liukiuensis.*
*S. liukiuensis*, Satake Palm (Ryukyu Islands). This palm has a handsome ringed trunk to 10 m (30′) high with a short green crownshaft. The 12, or more, feather leaves are 3 m (10′) in length and are neatly arranged in an attractive head above the crownshaft. The several pink inflorescence are stiff and branch horizontally. The fruits are oval, 13mm (½″) long, and black at maturity.
**Culture and habitat** *Satakentia* is a tall palm and in its natural environment may attain 15–20m (50′–60′) in height, standing taller than the other vegetation.

It is wind-hardy and withstands cyclones in its habitat. Although a magnificent palm, it is so far little grown. The few observed in the USA suggest that this palm could be quite hardy in tropical, subtropical and temperate gardens. It requires ample watering and seems to prefer to grow in the shelter of other vegetation in its early stages. It is closely related to *Clinostigma*.
**Sex** Monoecious. Self pollinating for seed production.
**Position indoors** No information is available.
**Propagation** Propagation is from seeds, but no information is available on germination times.

## SCHEELEA

28 solitary species distributed from Central America through to Brazil, Bolivia and Peru, growing in open grasslands and tropical rainforests.
*S. liebmannii* (Mexico). This tall robust species has a heavy trunk and long, rather upright feather leaves. The inflorescences hang below the leaves and the oval fruits are green to brown and 3-4cm (1″–1½″) long.
**Culture and habitat** *Scheeleas, Attaleas* and *Orbignyas* are all large palms mostly seen only in their natural habitat or in parks. Their habits and cultural needs are all very similar and are more fully dealt with under *Orbignya*.
**Sex** Monoecious. Self pollinating for seed production.
**Position indoors** Not known to be used indoors.
**Propagation** Fresh seed germinates within 2 months.

## SCHIPPIA

One solitary species from Belize, growing in the understorey of tropical rainforests.
*S. concolor*, Pimento Palm (Belize). This palm has a slender trunk to 10m (30′) high. There are about 12 deep shiny green fan leaves, split to the base, with slender stalks. Inflorescences extend beyond the leaves, and later bear round white fruits 25 mm (1″) in diameter.
**Culture and habitat** *Schippia* is a small fan palm that will stand some sun but should be given shelter in very hot areas. Being an understorey dweller, it is most suited to sheltered but warm sites. It requires a tropical to subtropical climate free from frost or cold winds, and should receive regular waterings. *Schippia* is a dainty plant and does not take up much room. It is somewhat similar to *Thrinax*.
**Sex** Hermaphrodite. Self pollinating for seed production.
**Position indoors** No knowledge of indoor use.
**Propagation** No precise information on seed germination times.

## SCLEROSPERMA

Three species, from west Africa.
Short clustering trunks and large almost undivided leaf.

*Schippia concolor*
The Pimento Palm of Belize.

*Scheelea liebmannii*
This palm is widely used by the local people in its Mexican homeland.

*Serenoa repens*
Saw Palmetto.
(Photo C. Migliaccio)

*Siphokentia beguinii*
This palm is popular with collectors.

## SERENOA

One clustered species, from the south-eastern USA, growing along the coastal sand dunes, in prairies and in pinelands.

*S. repens* (syn. *Sabal serrulata*), Saw Palmetto (south-eastern USA). A low-spreading, strong palm with subterranean or spreading trunks woven with brown fibre. These may eventually spread to 3m (10′) and form a wide heavy mass. The 1m (3′) fan leaves are deeply segmented and may vary in colour from yellowish-green to blue-green; they have a waxy texture. The branched inflorescence bears white flowers and the edible 1–1.5 cm (⅜″–⅝″) fruits are bluish to black.

**Culture and habitat** *Serenoa* is a hardy grower and will even tolerate some frost, as well as coastal winds. Its cultural requirements are somewhat similar to those of the more upright *Sabal* species, which are neighbours in its natural habitat. Refer there for additional information.

**Sex** Hermaphrodite. Self pollinating for seed production.

**Position indoors** No information available on use indoors.

**Propagation** Seed germinates regularly within 3–6 months. Growth is slow. Established plants do not transplant well.

## SIPHOKENTIA

Two solitary species, on islands in the north-western Moluccas, and confined to lowland rainforests.

*S. beguinii* (Moluccas). This shade-loving palm has a slender green trunk with prominent rings and a crownshaft, topped off with an umbrella of 2m (6′) feather leaves on which the leaflet tips remain unsplit. The pure white branched flower stalk is another attractive feature of this equatorial collector's palm; the 12mm (½″) fruits are reddish.

**Culture and habitat** Being equatorial, these palms would require a warm tropical climate if planted outdoors, as well as shade covering. In cooler climates a glasshouse may be best. There is a good specimen in the tropical dome at Brisbane Botanic Gardens, but because they are rarities, there is little stock to observe for further information, or to raise for commercial sale. This genus is very closely related to *Gronophyllum*; see there for further cultural advice.

**Sex** Monoecious. Self pollinating for seed production.

**Position indoors** No information available on indoor use.

**Propagation** Propagation is from seed. No germination times are available.

## SOCRATEA

12 solitary species (or possibly only five), from Nicaragua through to Bolivia and Brazil, growing in lowland and mountain rainforests.

*Socratea* sp. (juvenile)
A Stilt Palm from Bolivia.

*Syagrus comosa*
This is a slim species.

*Socratea exorrhiza*
Stilt Palm (centre), in the Rare Plants House, at Fairchild Gardens, Florida, USA.

*Syagrus* x *fairchildensis*
A *Syagrus Butia* seed cross, at Fairchild Gardens, Florida, USA.

*Syagrus flexuosa*
A species with multiple trunks.

*S. exorrhiza*, Stilt Palm (Brazil). This palm grows in the Rare Plants House at the Fairchild Gardens, Florida. It is a tall, slender plant supported by a framework of stilt roots, with a sparse head of feather leaves 1–2m (3′–6′) long with short stalks. The oval fruits are 26mm (1″) long.
*S.* sp., Stilt Palm (Bolivia). This interesting palm caught my eye when I was looking over a juvenile new collection of palm seedlings at the Royal Botanic Gardens, Sydney, grown from seeds received from Bogor Gardens in Indonesia.
**Culture and habitat** *Socrateas* are a rarity, along with their close relatives in the genus *Iriartea*. They are eye-catching because of the way the trunk is held above ground by the stilt roots at the base. On a fully matured palm, the trunk may be 3m (10′) off the ground, allowing a person to stand under it.

They grow tall, to about 20m (60′) with the trunk held above ground and upright on stilt roots, resembling *Verschaffeltia*. These stilt roots may even be seen on the juvenile plant which will lean over until the new root lifts it back upright again. The ringed trunk is topped with a bright green crownshaft and a sparse head of feather leaves with blunt ends. The leaflets stand out at different angles giving a plumose appearance. Small flower stalks appear from below the crownshaft, followed by oval 25 mm (1″) fruits, usually reddish-brown.
*Socratea* species are little cultivated, and then usually only in tropical botanical gardens that have the right conditions of humidity, high temperature and regular

*Syagrus picrophylla*
This species resembles *S. romanzoffiana.*

*Syagrus orinocensis*
A species of lighter stature.

moisture. Since they grow naturally in rainforests, extra shelter should also be beneficial.
**Sex** Monoecious. Self pollinating for seed production.
**Position indoors** As far as I know they are not used indoors.
**Propagation** Reports indicate that seeds germinate within 2 months. I should think that bottom heat would be required.

## SOMMIERIA

Three species, from New Guinea and Irian Jaya. Low solitary trunk, bearing stems of feathered leaves of only 2–4 leaflets.

## SYAGRUS

About 32 solitary or clustered species, now including *Arecastrum, Arikuryoba* and *Rhyticocos* (see Uhl and Dransfield, *Genera Palmarum*). They grow in South America from Venezuela to Argentina, with most in Brazil, and there is one species from the Lesser Antilles. Most species grow in dry areas, a few in tropical rainforests.
*S. amara* (syn. *Rhyticocos amara*), Overtop Palm, Lesser Antilles. Related to and resembling the Coconut (*Cocos*), this palm has a long slender trunk to 20m (60′) topped with a bushy head of 2–3m (6′–10′) feather fronds with green stalks 1–2m (3′–6′) long. The inflorescence emerges from a boat-shaped 30cm (12″) spathe amongst the leaves, and many inflorescences hang from the palm for long periods. The fruits are oval, 5cm (2″) long and orange at maturity. They contain a bitter milk when young and are hard when mature.
*S. comosa* (Brazil). This species very closely resembles *S. romanzoffiana*, but has wider leaflets on shorter stalks. The 25mm (1″) fruits are in smaller bunches, and are cream flecked brown.
*S. coronata* (Brazil), Licury Palm. The thicker 25cm (10″) trunk of this species, growing to 9m (30′), is covered in persistent old leaf bases. There is a thick head of about 25 feather leaves 3m (10′) long with 60cm (24″) stalks and these are held upright. A 90mm (3′) flower cluster is succeeded by orange fruits 25mm (1″) long.
*S.* x *fairchildensis* (at Fairchild Gardens). This is a seed cross between *Butia* and *Syagrus* and is an attractive palm with long recurving leaves. The long, even, bright green leaflets add to its attraction. No further details are available on fruiting habits.
*S. flexuosa* (Brazil). This palm has ringed, wrinkled trunks, wider at the base. The feather leaves are thin with narrow leaflets, but are held upright.
*S. orinocensis* (Venezuela, Honduras). Somewhat similar to *S. romanzoffiana*, this species is, however, lighter in stature and has smaller leaves with broader leaflets.
*S. picrophylla* (Brazil). This species has a ringed trunk and very much resembles *S. romanzoffiana*, but its

*Syagrus romanzoffiana*
Queen Palm.

*Syagrus romanzoffiana* var. *australis*
Showing the difference in fruits: the larger *S. r. australis* is on the left.

*Syagrus sancona*
A much taller grower.

*Syagrus schizophylla*
This is good as a potted palm.

inflorescences are smaller. The fruits are 25mm (1″) by 20mm (¾″) and yellow in colour.
*S. romanzoffiana* (syn. *Arecastrum romanzoffianum, Cocos plumosa*), Queen Palm (Brazil). This familiar feather palm is one of the most widely grown in Australia, being seen in gardens from Cairns to Melbourne. It has a grey trunk to 15m (50′) high, minus a crownshaft, crowned with a deep green, dense head of plume-like leaves which shine in the sun and move in the breeze. The spray of cream flowers is soon followed by large bunches of orange fruits 20–25mm (¾″–1″) long. If not removed early, these fall all over the ground.
*S. romanzoffiana* var. *australis* is much sturdier in its trunk and general appearance. It has larger oval fruits 30mm (1⅛″) long.
*S. sancona* (syn. *S. tessmanii*) (Peru). This is a tall palm to 20m (60′) much resembling the more widely used *S. romanzoffiana*. The trunk is enlarged at its base, showing rings, and is crowned by a head of feathery leaves. The 1m (3′) long inflorescence is branched, with yellow flowers followed by yellow fruits.
*S. schizophylla* (syn. *Arikuryoba schizophylla*) (Brazil). A species not so commonly grown. It has a single trunk about 2–3m (6′–10′) high, covered in erect old leaf bases which are purplish black. The dark green feather leaves arch gracefully. The fruit is orange-yellow and oblong, 32mm (1¼″) long.
*S. vagans* (Brazil). This species forms into a dense clump from a subterranean trunk. The deep green feather leaves are 2–3m (6′–10′) long and have narrow leaflets.
*S.* sp. (at Fairchild Gardens). This unusual species has a sturdy trunk which has branched 1m (3′) from the ground. It is carrying fruits, but no other information is available on this freak.

**Culture and habitat** Natives of Brazil, these adaptable palms are hardy in the tropical, subtropical and temperate areas of the world. They are suitable for seaside plantings and will also grow in inland areas, provided they receive an ample supply of water.

Because they do not develop their split leaves as small plants, *Syagrus* species are generally sold in larger-sized pots, 250–300mm (10″–12″) or more. They should be repotted if the soil is drying out too rapidly, usually one size larger at a time. A general potting mix is most suitable for potted specimens. If you are planting outdoors, potting mix may also be dug into 1–2m (3′–6′) of the natural soil.

**In the garden** These palms may be grown as solitary specimens or in clumps. They can also be used as large tub specimens where tall greenery is required. They withstand transplanting well and are widely used by landscapers in 'instant' gardens. (Very large, super-advanced plants are often successfully transplated in this way.) They are hardy in full sun, but will also tolerate no sunlight in warm climates. Watering should be kept up.

When fed regularly with a complete balanced fertiliser, the leaves are greener and broader and more attractive, and the trunk is also broader. They are very hardy, once established, but better growth results from regular watering.

**Sex** Monoecious. Self pollinating for seed production.
**Position indoors** Indoor use of these magnificent palms is restricted by ceiling heights, and the fact that the juvenile leaves do not split early. When they do, the plants immediately shoot up to heights of 2–3m (6′–10′) even in full sun. They are very hardy in high-ceilinged buildings, but last longer where natural light is available. Multiple plantings are also beneficial to promote a thicker plant. Never allow to dry out excessively indoors.
**Propagation** Propagation is from seeds, which start germinating in about 2 months. Grow the potted seedlings in full sunlight for a squatter growth.

## SYNECHANTHUS

Two to three species, from Mexico to north-west South America.
Solitary or clustered trunks, with feathered leaves.

## TECTIPHIALA

One species, *T. ferox*, from Mauritius.
Solitary or clustered spiny trunks, with feathered leaves.

## THRINAX

Seven solitary species, from the Caribbean area, found from sea-level to 1200m (4000′) in alkaline soils or sand, or around limestone outcrops, where they thrive exposed to salt winds. Also in dry evergreen woodlands and mountain rainforests.
*T. morrisii*, Peaberry Palm, Brittle Thatch Palm (Caribbean area). This species has a smooth, ringed 10m (30′) trunk, swollen at the base and with a clump of basal roots. The head of fan leaves is rather open and the leaves green above, silver beneath. Inflorescences are erect and arching, but later hang down when the small white fruits develop. This specimen was listed as *Coccothrinax morrisii* at Foster Gardens, Hawaii.
*T. parviflora* subsp. *parviflora*, Broom-thatch (Jamaica). The smooth trunk can vary between 3m (10′) and 8m (25′) tall. The fan leaves are almost circular, and the tips more twisted, and fold inwards. The round 8mm (5/16″) fruits are white.
*T. parviflora* subsp. *puberula*, (Jamaica). In this ssp. the head of fan leaves is dense, the leaf segments broad and leaf-to-stem joint prominent.
*T. radiata*, Florida Thatch Palm (Caribbean area). This is a hardy sea-coast fan palm with a trunk 3–8m (10′–25′) high. The leaves are deep green and the stalks have fibre at their bases. The erect inflores-

*Syagrus vagans*
This palm forms a thick clump.

*Syagrus* sp.
An oddity at Fairchild Gardens, Florida, USA. This specimen shows multiple branching high up on the trunk.

*Syagrus coronata*
Licury Palm has a thick trunk.

*Syagrus amara*
Overtop Palm is salt-tolerant.

cences carry white flowers and white fruits, 8mm (5/16″) in size.

**Culture and habitat** These are warmth-loving palms suited to tropical, subtropical and temperate zones. They will tolerate salt-laden winds, but should be protected from winds that are too cold. These small fan palms with slender trunks are very similar to *Coccothrinax*. The main noticeable differences are the mostly green leaves, the split leaf bases, the white fruits and the solitary trunk. Both genera are from similar regions on limestone soils, a point to remember when considering their soil requirements. They are also adaptable to dry areas with frost, but still prefer regular waterings. They will grow in isolated positions or amongst other vegetation, provided they have protection from cold winds. They can also be grown as sun-loving potted palms.
**Sex** Hermaphrodite. Self pollinating for seed production.
**Position indoors** Not known to be used indoors.
**Propagation** Seeds germinate within 3 months.

## TRACHYCARPUS

Six solitary or clustering species, occurring from the Himalayas of northern India to northern Thailand and

*Thrinax morrisii*
Peaberry Palm or Brittle Thatch Palm.

China, in temperate and subtropical regions up to 2400m (7500′) altitude. These are the most cold-tolerant of all palms.

*T. fortunei*, Chinese Windmill Palm (China). This species has a slender, messy trunk 3–12m (10′–40′) high covered with dark brown fibres and old leaf bases. The leaves are fan-shaped, with long divided segments, deep green above and greyish beneath. The inflorescences are small, with yellow, fragrant flowers and kidney-shaped bluish fruits 12mm (½″) long. Hardy in the British Isles and similar climates.

*T. martianus* (Himalayas, India). This palm has a trunk to 15m (50′) high, half bare, half covered with old leaf bases. The large, attractive fan leaves, half divided, have thin stalks 1.5m (5′) long. The inflorescences are followed by orange, to bluish fruits 12mm (½″) long. This species has been found growing at an altitude of 2400m (7500′).

*T. takil* (western Himalayas). This species is reported to grow in damp oak forests, with snow from November to March. It is of compact habit and slow-growing, with a shorter trunk than *T. fortunei*, to about 6m (20′) and covered in brown fur. There are abour 30 to 40 dark green fan leaves with narrow segments. The branched inflorescences have white flowers; fruits are kidney-shaped, 12mm x 9mm (½″x⅜″) and purplish. Said by some to be synonymous with *T. wagnerianus*

*T. wagnerianus* (from W. Himalayas) (at Huntingdon Gardens). This tall-growing species is over 10m (30′)

*Thrinax parviflora* ssp. *parviflora*
Broom Thatch. The leaf segments fold back under.

*Thrinax parviflora* ssp. *puberula* Thatch Palm. The leaves are almost circular.

*Thrinax radiata*
Florida Thatch Palm.

*Trachycarpus fortunei*
Chinese Windmill Palm withstands cold and light snow.

*Trachycarpus martianus*
This palm reaches an altitude of 2400m (7500′) in the Himalayas.

high, with a trunk covered in brown fibre. The stiff fan leaves are less than 1m (3′) long. There is a resemblance to *T. takil* (and some say they are one and the same palm) in the leaf shape, but *T. takil* is reported to be less than 6m (20′) high and to have broader leaf segments.

**Culture and habitat** *Trachycarpus* species grow in cold mountain regions of China and India and are hardy enough to withstand light snow. In horticulture they are mostly grown in the cooler regions, it being too warm in the tropics. They are often mistaken for *Chamaerops humilis*, the Mediterranean Fan Palm, but differ in many obvious ways and have a single trunk, as opposed to multiple trunks. They will grow in countries too cold for any other outdoor palms. They are also ideal for pots, where they can be moved around to suit the seasons or the occasion. Water well until fully established and shelter from hot sun, selecting positions away from hot walls and suntraps.

Pot on as required, using an indoor potting mix. Fertilise occasionally to maintain nice growth. These palms are often sold in a 200 mm (8″) pot, in which they are quite pleasing. The most attractive-sized plant is one growing in a 200–400mm (8″–16″) pot. They are best around poolsides, verandas and cool corners, also indoors for short displays.
**Sex** Monoecious. Self pollinating for seed production.
**Position indoors** Use only for short-term indoor display, then return to a veranda. Select a light and airy position. Maintain ample moisture.
**Propagation** Seed germinates readily within 2 months.

## TRITHRINAX

Five mainly solitary species distributed from Bolivia to southern Brazil and Argentina and growing from sandy marshes and along river banks, to dry areas, seeming to prefer alkaline soils of pH 7–8.
*T. acanthocoma*, Spiny Fibre Palm (Brazil). This palm has a thick trunk to 4m (12′) high, covered with brown spiny webbing and old leaf bases. The fan leaves, green above and paler beneath, are narrowly segmented. The heavy inflorescence has creamy flowers and round creamy white fruits 2cm (¾″) in diameter.
*T. campestris* (Argentina). This species forms a compact clump of trunks massed with attractive blue-green fan leaves that are fuzzy white on the upper surface. The showy fruits are 20–25mm (¾″–1″) long.

**Culture and habitat** These hardy palms require full sun and ample moisture. They are not generally seen very much commercially or outside botanic gardens, perhaps because of their spiky and untidy trunks. They seem quite hardy, with specimens growing in the tropical, subtropical and warm temperate zones.

Select a warm position away from cold winds, in soil which has been limed to a pH of 7–8 and supply good drainage.
**Sex** Hermaphrodite. Self pollinating for seed production.
**Position indoors** No knowledge of use indoors.
**Propagation** Seeds usually germinate in 3 months but can take longer.

## VEILLONIA

One species, *V. alba*, from New Caledonia.
Solitary trunk, feathered leaf palm.

## VEITCHIA

18 solitary species, from Fiji, New Hebrides, New Caledonia and the Philippines and growing from sea-level to 1000m (3000′) in forests.

*V. arecina* (New Hebrides). An attractive species with a large branched inflorescence and red fruits 2.5–5mm (1″–2″) long. It grows 6–9m (20′–30′) tall; the trunk is topped with a silvery crownhaft and leaves are 2.5m (8′) long.

*V. joannis* (Fiji). A tall feather palm having a grey trunk to 15m (50′) high topped with a long green crownshaft and dainty deep green feather leaves. The inflorescence is bunched and flowers are white; fruits are red and 4–5cm (1½″–2″) long.

*V. mcdanielsii* (New Hebrides). This is a slender feather palm to 8m (25′) high with a whitish ringed trunk topped with a grey-green crownshaft. There is a nice head of green feather leaves 3m (10′) long with grey-green stalks. A branched inflorescence of white flowers is followed by bright red fruits 3 cm (1¼″) long.

*V. merrillii*, Manila Palm, Christmas Palm (Philippines). This shorter species has a grey trunk to 7m (22′) topped with a green crownshaft. The deep green feather leaves recurve daintily. The stiff, bunched inflorescence of whitish flowers is followed by an attractive bunch of red fruits 2–3 cm (¾″–1¼″) long. This species is widely grown in warm climates for its display of red fruits.

*V. montgomeryana* (New Hebrides). This is best described as smaller version of *V. joannis*, similar in all features but only growing 8–12m (25′–40′) high. Fruits are 4cm (1½″).

*V. sessilifolia* (Fiji). This species has a brownish trunk to 11m (35′) high and has a brownish-green crownshaft and leaf stalks. The feather leaves are long 3m (10′) and arching. The numerous thick multibranched inflorescences, about 1m (3′) long, produce oblong reddish-orange fruits 2cm (¾″) long.

*Trachycarpus takil*
A shorter and slower-growing species.

*Trachycarpus wagnerianus*
This group has one specimen over 10m (30′)

*Trithrinax acanthocoma*
Spiny Fibre Palm. The name refers to the trunk.

*Veitchia merrillii* (centre)
In a setting with *Pritchardias.*

*Trithrinax campestris*
This palm forms a thick clump.

*Veitchia mcdanielsii*
The leaves of this species are held high.

*V. winin* (New Hebrides). This is a tall, slender feather palm to 20m (60′) high. The brownish trunk is topped with a pale crownshaft and brownish leaf stalks. The spreading inflorescence has creamy flowers and red fruits 2cm (¾″) long.

**Culture and habitat** These very clean, attractive, upright-growing palms are among the most popular palms in tropical regions everywhere. The bunches of red fruits are prolific and attractive.
*Veitchia* species come from warm tropical islands, growing in the protection of rainforest (the taller one eventually rise above the trees). They have adapted well to our tropical areas and handsome specimens may be seen in subtropical regions also.
*Veitchias* should be planted in warm, sheltered positions where there are no cold winds. They could be tried in the warm temperate zone as well, provided the area is warm, sheltered and frost-free and has mild winters. Here they may be best grown in the protection of other vegetation, but in the warmer zones they are widely used as single or clump specimens or line plantings both domestically and in the parks and botanic gardens. Water often.

**Sex** Monoecious. Self pollinating for seed production.
**Position indoors** *Veitchias* may also be used as indoor palms, and tub specimens for protected areas. *V. merrillii* is the best for this purpose, being fairly compact. In most countries *Veitchias* have been relatively unused indoors, with the exception of Hawaii, where they are frequently seen, being grown on until the attractive fruits are displayed.
**Propagation** Seeds germinate quickly (about 1 month), but plants should be protected from cold.

*Veitchia montgomeryana*
Has been described as a shorter version of *V. joannis.*

## VERSCHAFFELTIA

One solitary species from The Seychelles, growing on steep forested hillsides at altitudes from 300m–600m (1000′–2000′) usually singly.
*V. splendida,* Stilt Palm (Seychelles). Once seen, never forgotten. The dark trunk of this unusual and beautiful feather palm, which eventually may reach 20m (60′) grows on stilt-like roots. It is topped with a small crownshaft and beautiful dark green recurving leaves. On young plants the trunk and leaf stalks have prickles, but these disappear at maturity. The 1–2m (3′–6′) inflorescence bears yellow flowers and then brownish green fruits 25mm (1″) long.

**Culture and habitat** This eye-catching palm grows on elevated rocky hillsides in the protection of rainforests. It grows well out of doors in the tropics, but in cooler climates seems to be confined to glasshouses and conservatories in botanic gardens. Perhaps, if specimens were more plentiful, more could be learned about this interesting palm.

*Veitchia sessilifolia*
The crownshaft and stems are brownish.

*Veitchia winin*
This species has a creamy crownshaft and brownish stems.

*Veitchia joannis*
The red fruits are larger than in *V. arecina.*

*Verschaffeltia* would be best confined to tropical areas for outdoor plantings; in the subtropical fringe it could be planted in a sheltered position to guard against cool winds. In cooler areas it requires glasshouse protection where it is ideal in pots. It has even been used indoors. Potted specimens require regular watering.

**Sex** Monoecious. Self pollinating for seed production.
**Position indoors** Reports indicate that it is used in its younger stages in pots and tubs indoors, being very attractive. They would require a warm and humid situation with good natural lighting.
**Propagation** Given heat and humidity, seeds germinate in 2–3 months.

## VONITRA

Four species, from Madagascar.
Solitary or clustered trunks, with feathered leaves.

## WALLICHIA

Seven solitary or clustered species, from India to China to southern Thailand, growing in the understorey of tropical rainforests, from sea-level to 2000m (6000′).

*W. densiflora,* Wallich Palm (Himalayas, Assam). This is a palm without noticeable trunks, clustering from the base to a height of 2–3m (6′–10′). It has *Caryota*-type leaflets deep green above and whitish beneath. The inflorescence is also borne from the base of the plant, but is shorter than the leaves; the purple and yellow flowers are followed by small 12mm (½″) purple fruits.
*W. siamensis* (Thailand) Small clustering trunk variety with feathered leaves bearing *Caryota*-type leaflets.

**Culture and habitat** *Wallichias* originate from humid tropical forests ranging from sea level 2000m (6000′), with one species (*W. disticha*), growing on steep Sikkim sandstone declines and valleys.

They will tolerate some cold, growing outdoors in Sydney (warm temperate) and Brisbane (subtropical); they also thrive in the tropical climate of Cairns. Many people might not recognise some as being palms, as the plants more resemble a clump of large ferns.

When planting a *Wallichia* in the garden, select a protected, shaded area or a garden rainforest (cool in the tropics, warm in temperate areas). Make sure drainage is good, but maintain ample moisture.

**Sex** Monoecious. Self pollinating for seed production.
**Position indoors** No information available as to whether they are used indoors.
**Propagation** Propagation is from seeds, which are slow to germinate, and sometimes erratic. Young plants are cold-sensitive.

## WASHINGTONIA

Two solitary species, from south-western USA and north-western Mexico, in desert areas where there is ground and surface moisture.
*W. filifera*, Petticoat Palm, American Cotton Palm (south-western USA and north-west Mexico [Baja California]). This robust, hardy fan palm has a thick grey 15m (50′) trunk, fully covered with old leaves giving a skirt effect. The grey-green shiny leaves have threads attached and the green stalks have powerful thorns. White flowers are followed by oval brownish black fruits 10mm (3⁄8″) long. A species from desert regions, it tolerates dryness, heat and frost.
*W. robusta*, Cotton Palm, Washington Palm (north-western Mexico). This species has a brown trunk to 25m (80′) to which many old leaves remain attached. The fan leaves are green and shiny, and have orange stalks armed with vicious thorns. The young leaves have white cotton attached. The creamy flowers are followed by oval, brownish-black fruits 10mm (3⁄8″) long.

**Culture and habitat** *Washingtonias* are sturdy palms and are widely used through southern USA in street and specimen plantings. They can withstand extremes of temperature and moisture. Since their native climate is similar to that in many parts of Australia, they are also gaining popularity there. *W. robusta* is the one most used in Australia, although it has been found in the USA that *W. filifera* will handle climatic extremes better, while *W. robusta* is better confined to the milder coastal regions.

When planting, select a warm, open, sunny, well-drained position, with plenty of room. I once saw two young *Washingtonias* planted on either side of a narrow drive entrance. The owners would have been very sorry people in a year or two when they were forced to remove two very handsome palms. Keep them away from traffic areas, as the spikes are very vicious if run into. In dry areas *W. robusta* would need a moister position. Unlike *Livistonas*, *Washingtonias* will take full sun from an early age. If you are moving specimens growing on open ground for landscaping, it is best to remove all except the youngest four leaves from the trunk, to prevent deterioration.

*Washingtonias* are ideal palms for tubs in open, sunny areas. Water regularly to keep a good appearance.

*Veitchia arecina*
The large branched inflorescence is now carrying fruits.

*Verschaffeltia splendida*
This Stilt Palm prefers the tropics.

*Wallichia siamensis*
Small palm on clustered trunks.

*Wallichia densiflora*
Wallich Palm forms a clump.

**Sex** Hermaphrodite. Self pollinating for seed production.
**Position indoors** I have not found them a good palm for indoor use: they have suffered from tip burn. However, some commercial customers have reported them to be good in plant hire. If you try them indoors, observe them regularly.
**Propagation** Seeds germinate easily and within 2 months. Deep pots are required to handle the deep roots of the young plants.

## WELFIA

Two species, from Honduras and Colombia.
Solitary trunks and feathered leaves.

## WENDLANDIELLA

Three species, from the Amazon basin of Peru.
Dwarf clustering thin trunks, with leaves of 2–4 leaflets.

*Washingtonia filifera*
American Cotton Palm or Petticoat Palm.

## WETTINIA

Nine solitary or clustered species, from Panama and north-western South America, growing as stilt palms in very wet rainforests at low to medium elevations.
*W. fascicularis* Stilt Palm (Colombia). This species has a long 11–12m (35′–40′) slender, ringed trunk, with a crownshaft topped with a sparse head of feather leaves. The base of the trunk has stilt roots, covered in small sharp lateral roots. The unusual inflorescence is branched and crowded, emerging from a sheath, and bears cone-like bunches of unusual blocky, crowded fawn fruits, 14mm (½″) long, covered in a spiky fuzz.

**Culture and habitat** *Wettinia* species occur in wet areas, or near streams, at low to medium elevations in the forests of Peru and Colombia. They are dainty medium-sized palms to about 12m (40′) high. It is stated that as they emerge above the tree canopy, their feather leaves droop downwards, as if to avoid the hot sunshine. They grow alongside stands of *Iriartea* which likewise have stilt roots. The flowers are borne on unusual, tassel-like hanging stems, while the fruits hang in sausage-like bunches reminiscent of cycad cones.

The locals use the leaves for thatching and the trunk timber is split and used for flooring and in walls. The above species is under trial in the Royal Botanic Gardens, Sydney, to gain further cultural knowledge for that area.

These palms are best grown in botanic gardens or by collectors, as they require conditions rarely available in private gardens. They would need an ever-wet, humid area in a rainforest or similar environment.
**Sex** Monoecious. Self pollinating for seed production.
**Position indoors** Not known to be used indoors.
**Propagation** No information is available on germination times, but it is thought that bottom heat and high humidity would be essential.

## WODYETIA

One solitary species, from north-eastern Australia, growing on the southern Melville Range, in open woodland and rainforest, as a canopy palm.
*W. bifurcata*, Foxtail Palm (north-eastern Queensland). This is an attractive palm, having a smooth, closely ringed trunk 10–15m (30′–50′) tall topped with a narrower green crownshaft covered in white bloom. The leaves have very narrow leaflets, arranged in a circular plume and much like those of a *Normanbya*.

*Washingtonia robusta (front)*
Washington Palm or Cotton Palm.

*Wettinia fascicularis* (juvenile)
This Stilt Palm comes from Colombia.

*Wodyetia bifurcata*
Foxtail Palm comes from north-eastern Queensland.

*Zombia antillarum*
Zombie Palm has unusually arranged trunk spines.

The green inflorescence is borne on a thick stem emerging from below the crownshaft, and the oval fruits are large 3.5 x 5.5cm (1⅜" x 2¼") and attractively displayed, turning orange-red at maturity.

**Culture and habitat** This most attractive Australian palm has been little known until very recent years, even in Australia. It was even absent from the numerous botanic gardens. This, however, is now changing with a developing world interest creating a demand for stocks. The name *Wodyetia* comes from that of a now deceased Aboriginal bushman, Wodyeti. This palm closely resembles its relative *Normanbya*. *Wodyetias* are to be found in rocky areas, open or bushy, in seasonally dry sandy soil, where they develop a deep root system early to enable them to survive between the monsoonal rains. Cultural information for this new palm is still being assessed, but it is thought that they would be frost-sensitive, but less cold-sensitive than *Normanbya*. The most advanced of the early planted specimens are growing well in Brisbane (subtropical) and Sydney (warm temperate). Their natural habitat is in hot sun, so they should be allowed to grow up into this, while being sheltered from cold winds. They would require a deep, well-drained soil, possibly sandy, and would probably withstand periodic dry spells, once properly established. Deep soakings would be the best way to apply water.
**Sex** Monoecious. Self pollinating for seed production.
**Position indoors** As yet untried.
**Propagation** Seeds begin germinating in about 2 months, but germination may be drawn out. They root deeply, so provision should be made for this. Pot into deep pots and hold plants in warm areas.

## ZOMBIA

One clustering species, from Hispaniola and growing on open bushy slopes and very dry hills.
*Z. antillarum,* Zombie Palm (Hispaniola). The multiple trunks 3m (10') high are covered in patterns of down-pointing brown spines. The fan leaves are bright green above and silvery beneath, with quite thin stalks. The 45cm (18") inflorescence is branched; the white spongy apple-shaped fruits are 25mm (1") long.

**Culture and habitat** *Zombia,* which originates from Haiti, is little grown out of botanic gardens, but is quite attractive when observed there. This is perhaps typical of thorny-trunked palms. It requires a sunny position. As it is closely related to both *Coccothrinax* and *Thrinax,* the growing advice for both those genera would also apply here.
**Sex** Monoecious. Self pollinating for seed production.
**Position indoors** Not known to be used indoors.
**Propagation** Seeds germinate in 2 months.

# CYCAD VARIETIES

The varieties are arranged alphabetically.

## AUTHOR'S NOTE

Some species of Cycads still require further research by botanists. When this has been carried out changes could occur in the naming, thus reducing or adding to the listings. Such research could be a long time away, depending on when priority can be given to the subject, and is likely to take much negotiation. That can be a very slow process because of the delays caused by the constant exchange of information, back and forth between world botanists.

The listing here is as they appear in the various botanical gardens, or from collectors, or nurseries. These have been looked over by a botanist. He has amended some, but will not verify any of which he may have doubts or is unable to research further. I have chosen to include these for a fuller listing of those varieties now in cultivation in various parts of the world.

In nature, cycads often occur in lonely, difficult to reach, politically unstable, or even militarily active places which may be unsafe to visitors. This hampers quick research and is a problem particularly relevant to *Ceratozamias, Dioons* and *Zamias,* from Central and South America.

Many variations can occur with cycads in cultivation, amongst collectors or botanical gardens, where only one or two of a variety are in that collection. Plants being singly sexed may be pollinated by a different variety, making variations which are very obvious, or not so obvious. In larger stands of a variety located in natural habitats this will be less likely to occur. This factor should be taken into account when viewing these pictures, or any plants you may also have obtained from friends or growers, no matter how reliable you may think they are.

*Cycas revoluta*
This huge aged specimen has formed a clump at Huntington Gardens, Pasadena, California.

# BOWENIA (ZAMIACEAE)

*Bowenia serrulata*
Note the fern-like fronds and habit of this cycad.

Two species, naturally growing in coastal regions of north-eastern Queensland. *B. serrulata* is found in thick stands in eucalypt forests, and *B. spectabilis* in rainforests, accompanied by tree ferns and other ferns.

*Bowenias* are regularly mistaken for ferns because of their appearance, habit and growth. They differ from ferns in the stiffness of their foliage. The cut fronds have long been used as decoration, and last longer than those of ferns. *Bowenias* can be distinguished from the other genera of cycads by the unusual shape of their branched fronds.

*B. serrulata,* Byfield Fern (central coast Queensland). This species has a thick subterranean base with seven to 10 branched fronds 1–1.5m (3′–5′) high. The deep green leaflets have serrated edges and are stiffer than those of ferns.

*B. spectabilis,* Zamia Fern (north coast Queensland). This is similar to the above species, with about seven fronds and having leaflets with smooth edges. This species is somewhat variable, resulting in a narrower frond form within the species; from the Mission Beach area near Cairns.

**Culture and habitat** *Bowenia* species would be best planted in sheltered areas in filtered sun or shade, and would need regular hosing. Rainforest or fernery conditions would be best. They are also good potted plants. The seed and pollen cones are hidden around

*Bowenia serrulata*
Byfield Fern.
(Photo L. Butt)

*Bowenia spectabilis*
*Zamia* Fern.

*Bowenia spectabilis*
A narrow-leaved form.
(Photo L. Butt)

*Ceratozamia mexicana*
Mexican Horncone. Popularly grown.

*Ceratozamia microstrobilis* (juvenile)
A juvenile plant, showing early form.

*Ceratozamia miqueliana*
The upright leafstems have thorns.

*Ceratozamia mexicana*
A juvenile plant, showing early form.

the base of the plant and are often only noticed when the seeds mature; they are greenish-brown when young and blackish-brown at maturity, with reddish-brown oval nuts.
**Sex** Dioecious. Need a male and female for fertile seeds.
**Position indoors** Little information as yet. See 'Cycads for indoor decoration' (p. 41). Good natural light would be an advantage.
**Propagation** The nuts are generally half buried in moist potting mix and kept damp. Also refer to the appendix, 'The Cycads of Australia' (p. 13). Divisions may also be taken from older plants; be careful to remove the soil from one side of the plant for clearer viewing, so as not to damage the balance of the plant.

## CERATOZAMIA (ZAMIACEAE)

About 17 species recorded, occurring from Mexico to Guatemala and growing as understorey plants.
*C. hildae* (Mexico). This species has multiple slender leafstems bearing groups of mid-green leaves.
*C. kuesteriana* (Mexico). Also multiple-stemmed, but with the leaflets longer and more closely arranged.
*C. latifolia* (Mexico). Multiple-stemmed, the leafstems arch outwards, with stiff tapered shiny leaflets.
*C. longifolia* (Mexico). Much resembling *C. mexicana,* but with the leaflets longer on mature plants.
*C. mexicana,* Mexican Horncone (Guatemala, Mex-

*Ceratozamia norstogii*
The new growths are silvery.

*Ceratozamia hildae*
The stems form an upright clump.

ico). Multiple spreading leafstems with long narrow leaflets, closely arranged.
*C. microstrobilis* (Mexico). Multiple leafstems bearing widely spaced, broad, tapered leaflets. Later the leaflets are longer and narrower.
*C. miqueliana* (Mexico). Multiple upright leafstems short leaflets are alternately arranged.
*C. norstogii* (Mexico). A small species having multiple thin leafstems and narrow blue-green leaflets. The new growths are coated with silver.
*C. 'Palma Solo'* (Is possibly *C. mexicana* var. *robusta*). Has multiple yellowish leafstems and bright green leaflets tapering to a point.
*C. plumosa* (Mexico). Has multiple floppy leafstems which bear narrow green leaflets twisting at different angles, giving a plumose appearance.
*C. purpusii* (Mexico). An old name and could be synonymous with *C. norstogii.* Multiple leafstems, but relatively few in number, bearing very narrow, long leaflets, closely arranged in a backwards V shape.
*C. recurvata* The validity of this variety is doubtful. An upright grower with an above-ground trunk. The tapered green leaflets grow horizontally and then recurve downwards.
*C. robusta* (Mexico). A sturdy variety resembling *C. mexicana* but more sturdy and upright in growth.
*C. 'Santiago Tuxla'* A recent introduction and possibly *C. mexicana* var. *latifolia* or *C. miqueliana.* No larger specimens were available for inspection.
*C.* sp. (Mexico). Multiple-stemmed arching species,

*Ceratozamia latifolia (mexicana latifolia)*
The fronds arch outwards.

*Ceratozamia longifolia* (syn. *C. mexicana*)
This plant has a cone.

*Ceratozamia kuesteriana*
This species spreads attractively
(Photo C. Migliaccio)

*Ceratozamia recurvata*
This species has a sturdy trunk.

*Ceratozamia* sp. Queretaro (juvenile)
The leaflets and stems are silver. See notes.

*Ceratozamia* sp. *'Santiago Tuxla'* (juvenile)
A recently introduced variety. See notes.

*Ceratozamia 'Palma Solo'* (juvenile)
A recent introduction, possibly *C. mexicana* var. *robusta.*

with drooping, very narrow, closely arranged deep green leaflets.
*C.* sp. *Queretaro* (Mexico) another recent introduction and possibly *C. latifolia* or *C. microstrobilis.* Once again no larger specimens were available for inspection. The narrow leaflets are silver beneath.
Other species not pictured:
*C. brevifrons* (syn. *C. mexicana*)
*C. matudae* (Mexico)
*C. nurangjo* (Mexico)
*C. sanuincente* (The species is doubtful)
*C. spinosa* (Syn. *C. robusta*)
*C. zaragozae* (Mexico)
*C.* sp. nov. (Mexico)
**Culture and habitat** Some *Ceratozamia* species have trunks as much as 2–3m (6′–10′) in height and 25cm (1″) in thickness, while others have much lower, or even subterranean, trunks. The leaf fronds are less symmetrically arranged than most of the other cycads, perhaps because their stalks are thinner and less stiff. Some do have a few spikes along the stalks but on the whole they are soft and less dangerous to handle.

Being understorey plants, they would burn if placed out in hot sun, cold winds or frosty conditions. They are best planted under the cover of trees or palms, in ferneries, rainforests, or other sheltered areas. They are also excellent as potted specimens, as this displays their features without detracting from other leaf forms around them. Potting gives the advantage of portability, allowing pots to be moved

*Ceratozamia plumosa*
A species with narrow leaflets.

*Ceratozamia purpusii* (syn. *C. norstogii*)
The leafstems are purple.

*Ceratozamia robusta*
The stiffer leaf fronds are widely displayed.

*Cycas armstrongii*
In natural rocky habitat, east of Darwin, Northern Territory.
(Photo D. Stallard)

to better locations so as to avoid any positional or climatic damage to the foliage. In pots use a rich, open potting mix, and water well and often.
**Sex** Dioecious. The female cone is broader than the male. Needs a male and female for fertile seeds.
**Position indoors** No detailed information is available on use indoors but, like other cycads, they are showing great promise. Select a position of highest natural light, keep moist (without water in the saucer) and remove outdoors to a sheltered protected position (greenhouse, veranda, etc.) to let the plant make its new leaves. When these are hardened off, the plant may be returned indoors again.
**Propagation** Fertile nuts (seeds) are half covered in potting mix and kept damp. They may be grouped at first, but when growth is first observed they should be planted singly to avoid root damage on moving. Safe-to-get-at divisions may be removed from older plants with care.

## CYCAS (CYCADACEAE)

About 52 established species, in northern Australia, South-East Asia, Japan and the Pacific Islands, with a few species from India, Madagascar and Africa. Most appear to grow in hot, seasonally dry, open areas. This is the most widespread genus of cycads.
*C. angulata* (Australia — Northern Territory). Light to mid-green fronds. Grows in eastern Arnhem Land and on coastal islands in eucalypt forests.

*Cycas cairnsiana*
With male cone.

*Cycas armstrongii*
Ideal as a pot-plant.

*C. armstrongii* (Northern Territory and Western Australia). Up to 60 mid-green flexible fronds. Grows eastward from Darwin and in the Kimberleys. Orange fruits. Moderate grower.
*C. basaltica* (Australia). This cycad grows on rocky hillsides in the Kimberley Ranges of Western Australia. The trunk bears angled to horizontal, stiff grey-green fronds.
*C. cairnsiana* (Queensland). Like *C. revoluta,* but with narrower leaflets on recurved fronds. Grows in areas inland from Cairns.
*C. calcicola* (Australia — Northern Territory). The flat fronds are inclined to be hairy. Grows on limestone outcrops near Katherine and south of Darwin.
*C. circinalis,* Fern Palm (India to Thailand and Sri Lanka). Has broad, stiff, glossy deep green fronds. The handsome leaflets are recurved. A large grower.
*C. circinalis* var. *orixansis* is found in India (not pictured).
*C. furfuracea* (Western Australia). This species comes from the Kimberley Ranges. It stands on a trunk topped with the grey-green leaflets angled upwards from the leafstem.
*C. kennedyana* (Queensland). The stiff deep green fronds have leaflets angled upright on the stems. Grows in the Normanby Range near Bowen, and appears to be a selection from *C. media.*
*C. lane-poolei* (Western Australia). Grows in eucalypt forests in northern Western Australia. It is possibly the same as *C. armstrongii,* and certainly appears very similar, although it has shorter leaflets.
*C. media,* Australian Nut Palm (Queensland). The mid

*Cycas basaltica*
The leaves are grey-green.

*Cycas angulata*
Ancient giants in natural habitat, Gulf of Carpentaria, Northern Territory. (Photo D. Stallard)

*Cycas angulata* (female plant)
In natural habitat in Arnhem Land, northern Australia.
(Photo D. Stallard)

*Cycas kennedyana*
A large plant with cones.

*Cycas lane-poolei*
This may be the same species as *C. armstrongii.*

*Cycas media*
Australian Nut Palm.

*Cycas neocaledonica*
Bread Palm, from New Caledonia.

to deep green fronds have long wavy leaflets. Grows from Rockhampton to Cape York in rainforests and eucalypt forests.

*C. neocaledonica*, Bread Palm (Australia?, Indonesia, New Caledonia). A vigorous species with long leaflets.

*C. normanbyana* (Queensland). Similar to *C. media* and possibly a localised variation of that variety. Leaf fronds 90–150cm (3′–5′) and leaflets 25cm (10″) long, mid to deep green. Fruits are orange. Grows in the Proserpine area and the ranges and estuary of the Burdekin River.

*C. pruinosa* (Western Australia). Shrubby, with grey-green fronds, bearing leaflets which are angled upwards. Grows in the Kimberleys (Deception Range area) on rocky hillsides. There are two distinct forms, one having bright green fronds and the other blue-green.

*C. pectinata* (South-East Asia). The deep green fronds are long and arching, with narrow leaflets closely arranged.

*C. revoluta*, Sago Palm (Ryukyu and nearby islands). The most popular and widely cultivated cycad. Fronds are a glossy deep green and with leaflets which are angled upwards. A number of varieties of this species have been described, including var. *binervis*, var. *robusta*, forma *planifolia* and *nana* (Shusta).

*C. rumphii* (eastern Asia and Pacific). A large species similar to *C. media*. The fronds are irregular and the leaflets narrow.

*C. siamensis* (South-East Asia). Lovely broad fronds

*Cycas furfuracea*
The leaves are grey-green.

*Cycas circinalis*
Fern Palm is a most attractive cycad.

*Cycas circinalis*
An aged specimen with many trunks

*Cycas pruinosa*
A green form from the Kimberleys, Western Australia.

*Cycas pruinosa*
A blue form from the Kimberleys, Western Australia.

*Cycas revoluta*
Sago Palm. A female plant showing seeds.

*Cycas media*
Male cone at front, female plant at the rear with seeds. In natural habitat, north-eastern Australia.

with silvery, later deep green narrow leaflets. Appears difficult to grow in some areas.
*C. taiwaniana* (China, Taiwan). The deep green fronds grow larger than those of *C. revoluta*.
*C. thouarsii* (south-eastern Africa, Madagascar). This juvenile plant had blue-green, semi-stiff leaflets.
*C. wadei* (Philippines). Fronds are broader and leaf-stems longer than in *C. revoluta*.
*C.* sp., Australia, growing at Foster Gardens, Hawaii. Similar to *C. media*, and appears to be that variety.
*C. 'Glen Idle Blue'* (Queensland). An attractive silvery blue species, growing near Glen Idle Station in north Queensland. Resembles *C. cairnsiana*, as a blue form.
*C.* sp., Pine Creek (Australia). An unusual species from north of Katherine in the Northern Territory. It differs from most other species of *Cycas* in the close spacing of the blue-green leaflets, as well as in the uprightness of the fronds. Resembles *C. armstrongii* as a juvenile.
*C.* sp. *FL.2589* (Fiji). A species growing at Foster Gardens, Hawaii.
*C.* sp. *TG.135 632369* (Guadalcanal). Growing at Foster Gardens, Hawaii.
*C.* sp. *Maryborough Blue* (Queensland). Blue-green fronds. Grows near Marlborough, north Queensland. Probably a form of *C. media*.
Other species are:
*C. balansae* (Vietnam)
*C. beddomei* (India)
*C. celebica* (Sulawesi)
*C. chamberlainii* (Philippines)

*Cycas media*
This plant is hundreds of years old. Growing in Cairns Botanical Gardens.

*Cycas pectinata*
This species has very narrow leaflets. (Growing at Foster Gardens, Hawaii.)

*Cycas normanbyana*
Very similar to C. media.

*Cycas* sp. Pine Creek (juvenile)
Perhaps a juvenile of *C. armstrongii,*
common in that area.

*Cycas wadei*
A cycad from the Philippines.

*Cycas pruinosa (male)*
Note the male cones, or pollen cones
(Photo L. Butt)

*Cycas pruinosa (female)*
Note the colourful fruits. This blue form has now been named Cycas "Surprise Blue". Found growing in the Mt. Suprise region of north-east Queensland. (Photo Irene Champion)

*C. chevalieri* (Vietnam)
*C. comoroensis* (Comore Islands)
*C. gracilis* (Australia) (possibly *C. armstrongii*)
*C. hainanensis* (China)
*C. jenkinsoniana* (origin?)
*C. madagascariensis* (Madagascar)
*C. micholitzii* (China, Vietnam)
*C. nathorstii* (Sri Lanka)
*C. panzhihuanensis* (China)
*C. papuana* (New Guinea and surrounding islands)
*C. repanda* (syn. possibly *C. micholitzii* var. *simplici pinna*) (Southern China)
*C. revoluta* sp. by Satake for acceptance as validated species are:— *cristata, diplofoliolum, C. corrugala, variegata, alba, aurea, hystrix, involuta* and *glabra.*
*C. riuminiana* (Philippines)
*C. rumphii* var. *bifida*
*C. rumphii* var. *bifida* ssp. *zeylanica*
*C. schumanniana* (Papua New Guinea)
*C. seemannii* (Fiji, Caroline Islands, New Caledonia)
*C. sphaerica*
*C. szechuanensis* (China)
*C.* sp. nov., WA (Western Australia)
*C.* sp. Foxworthy (Philippines)

**Culture and habitat** *Cycas* is a popular and large genus, naturally located in both northern and southern hemispheres. *Cycas* species have a single trunk, sometimes branched, and some of the species attain a height of up to 5m (15′). Large specimens may be many hundreds of years old. The black trunks are topped with a handsome display of stiff fronds; the

*Cycas* sp. FL.2589 (Fiji)
Growing at Foster Gardens, Hawaii.

*Cycas thouarsii* (juvenile)
species from South Africa and Madagascar.

*Cycas revoluta*
Sago Palm. A male plant with two cones.

*Cycas rumphii*
This cycad has a massive leaf spread.

*Cycas* sp. (Philippines)
Produced from seeds, sold commercially as *C. rumphii.*

*Cycas siamensis*
Apparently a sparse-leaved species.

*Cycas taiwaniana*
The leaflets of this female plant are quite narrow.

leaflets have sharp tips and are very glossy. The colour varies according to species, from silver and silvery blue to the deepest green, but even deep green species may have their juvenile fronds opening silvery at first. *Cycas* species differ from other cycads in the development of their female seed cones. They form a large cone in the centre of the plant, then from the centre of this, before the seeds have ripened, the next year's new fronds emerge. The male cones stand upright in the centre of the plant. *Cycas* species have common names like 'Sago Palm', 'Fern Palm', 'Nut Palm', 'Bread Palm' and 'Queen Sago'.

*Cycas revoluta* is the most popularly grown of all cycads, perhaps because it, like the palm *Rhapis*, was a favourite in early Chinese and Japanese horticulture. It is now available from nurseries through most of the world, where warmer climates prevail. There are also many other lovely species that will become just as popular once supply catches up.

*Cycas* species appear to be quite hardy in most garden positions, from full hot sun to total shade. They will grow in desert regions, and may tolerate light frosts. However, their foliage takes on a better appearance without climatic extremes, or damage from tree droppings. The sun puts a nice healthy, glossy stiffness into the leaves. In landscaping, *cycas* are very pleasing garden specimens. Slow-growing and long-lasting, compact plants, without any need for constant attention, they are relatively disease- and insect-free. Use them as specimen plants in the garden and around the swimming pool, but out of traffic areas.

*Cycas* sp. TG.135 632369, (Gaudalcanal)
Female plant, growing at Foster Gardens, Hawaii.

*Cycas* sp., (Australia), at Foster Gardens, Hawaii.
This female species resembles *C. media*.

*Cycas 'Marlborough Blue'*
The blue foliage matures blue-green.

*Cycas 'Glen Idle Blue'*
Attractive blue-green leaves, resembling *C. cairnsiana.*

*Cycas 'Glen Idle Blue'*
Originating near Glen Idle Station in North Queensland, this cycad is popular in Australia.

*Cycas 'Marlborough Blue'*
Located near Marlborough, Queensland. Probably a form of *C. media.*

*Dioon califanoi* (juvenile)
An attractive silver species.

**Sex** Dioecious. Need a male and female for fertile seeds.
**Position indoors** Select a position with good natural light to minimise leaf damage over an extended period indoors. Plant lights may be helpful if the lighting is dim. *C. circinalis* and *C. revoluta* have both given favourable results when tested for short periods indoors. Refer to 'Cycads for indoor decoration' (p. 41).
**Propagation** Fertile nuts are best half buried in potting mix and kept moist. See appendix 'The cycads of Australia'. Rooted offshoots from the base of an older plant may be carefully removed and planted. Cut trunks without roots may be propped up on sand until roots appear. Keep the stem and leaves sprayed daily with water to avoid dehydration, but do not let the sand become over-wet.

## DIOON (ZAMIACEAE)

11 species recorded, growing along the edges of Mexican rainforests and also in open sun, amongst bromeliads, cacti and ferns. They seem to survive well in dry conditions.
*D. califanoi* (Mexico). Juvenile growth silvery, mature growth blue-green, with silver scurf at base.
*D. edule*, Virgin's Palm (Honduras, Mexico). Juvenile fronds white-woolly. The 1–2m (3′–7′) long fronds are beautifully displayed and the leaflets have spiny tips. Edible seeds in *F. cones* 30 cm (12″) long.
*D. edule* var. *angustifolium* (Mexico). Fronds narrower than *D. edule*, and blue-green.

*Dioon 'El Cameronii'* (juvenile)
A recent introduction, from Mexico.

*Dioon edule* var. *angustifolium*
This one has narrower fronds.

*Dioon mejiae*
A species from Honduras.

***Dioon edule***
**Virgin's Palm with female cone.**

***Dioon purpusii***
**The new growth is covered with silver scurf.**

***Dioon edule*** **var.** ***nova***
**This variety has broader fronds. See notes.**

***Dioon tomasellii*** **var.** ***sonorense***
**This blue-leaved variety seems difficult to grow.**

*D. edule* var. *nova*. An undescribed variety with an appearance of broader, blue-green fronds. The authenticity of this variety is in doubt.
*D. 'El Cameronii'* (Oaxaca Mexico). The juvenile growth of this species is covered with white fur.
*D. mejiae* (Honduras). The fronds of this species are mid-green with yellow stems.
*D. merolae* (Chiapas, Mexico). The deep green fronds have reflexed leaflets and are narrower.
*D. purpusii* (Mexico). Juvenile growth silvery and woolly, becoming dark green with paler reverse. Leaflets have three side teeth. Male cone 20cm (8") long, female cone 45cm (18") woolly.
*D. spinulosum*, Giant Dioon (Mexico). This is the tallest of the species, with trunks to 10m (30') tall. It has attractive deep green fronds with leaflets edged with teeth. The female cone is acorn-like and pointed.
*D. tomasellii* (Mexico) was described in an article in *Brittonia* 36: 223–227 in 1984, as a new variety from Western Mexico with two varieties. One is:—
*D. tomasellii* var. *sonorense* (Mexican state of Sonora and north Sinaloa). The foliage on this variety is sparse and appears rather hard to grow. Its blue leaf colouring is very attractive.
Other recorded species are:
*D. caputoi* (Mexico)
*D. dohenyi* (Mexico)
*D. holmgrenii* (Mexico)
*D. rzedowskii* (Mexico)

**Culture and habitat** The name *Dioon* means 'two eggs' and hence should be pronounced di-o-on, from the fact that there are two seeds behind each scale on the female cone. These cones form singly in the central shoot or shoots of the plant. They are brownish and covered with silver fuzz and because of their great weight usually droop over.

Dioons grow in full sunlight. They are either tall and tree-fern-like with a single trunk to 10m (30') (e.g., *D. spinulosum*, a rainforest edge species), or shorter with multiple trunks. Their foliage is the most attractive and symmetrical to be found in any cycad. This symmetry makes them very attractive for display as a potted plant, but remember to place them out of traffic areas or they may cause injury. Dioons also adapt well to more shaded positions, but may get a buildup of moss on their fronds; in this case, clean the fronds off regularly.

**Sex** Dioecious. Needs a male and female for fertile seeds.

**Position indoors** Little information is available on their use indoors but, like other cycads, they are showing great promise as long-term indoor plants. Select a position with good natural light, keep moist (do not leave water in the saucer) and remove outdoors if they begin to make new leaves. Refer to general text.

**Propagation** As with other cycads, half cover the seeds in potting mix and keep moist. Replant singly at the first sign of shooting, to avoid root damage. They can also be propagated from divisions (offshoots) with roots, or from cut trunks. Refer to general text.

*Dioon merolae*
Note the pale leafstem, and the female cone.

*Dioon spinulosum*
Giant Dioon, with a female cone.

*Encephalartos bubalinus*
An upright-growing species.

*Encephalartos cupidus*
This species has low suckering growths.

*Encephalartos barteri*
The cones are green and the leaves have a twist.

*Encephalartos eugene-maraisii*
More colourful and hardier in full sunlight.

# ENCEPHALARTOS (ZAMIACEAE)

51 species recorded, confined to Africa. They are to be found on seashore sand dunes, in grass veldt, or in rocky or grassy, elevated areas amongst timber, even in regions receiving snow, to at least 1800m (6000′).

*E. altensteinii,* Prickly Cycad (East Cape, Transkei). A sturdy grower on a 4–5m (12′–16′) trunk, with the stiff 1–2m (3′–6′) fronds having narrow, fresh green leaflets with 2-3 spines. Multiple cones are golden-yellow, fruits red. Semi-hardy, but frost sensitive.

*E. arenarius,* (East Cape). Forms a clump, or trunk to 1m. (3′) stiff, recurving 1–1.5m (3′–5′) fronds, having dark green, forked and spined leaflets. Cone is light green, fruits shiny red. Semi-hardy, but no frost, good in sand.

*E. barteri* (Benin, Ghana, Nigeria). A less stiff variety, from humid tropical regions, with a lighter trunk and wavy, spreading fronds 1.5–2m (5′–6′) and narrow, green leaflets. Cones are green. No frost, suits some shade. There is a subspecies from Nigeria, *E. barteri* ssp. *allochrous,* but no information is available.

*E. bubalinus* (Tanzania, Kenya). From the trunk there is an upright display of 1–2m (3′–6′) stiff fronds, with narrow, bright green leaflets. Semi-hardy but no frost.

*E. caffer,* Kaffir Bread. (East Cape). From the subterranean trunk are borne 0.5–1m (18″–3′). Woolly fronds, bearing densely arranged, narrow, green and woolly leaflets. Multiple cones are yellow. Sun hardy, no frost, prefers acid soil.

*Encephalartos altensteinii*
The Prickly Cycad grows in sun or shade. This female plant has two cones.

*Encephalartos 'Chinhazany'*
The fronds of this plant have arching yellow stalks.

*Encephalartos arenarius*
This species suits sandy soils.

*Encephalartos gratus*
This cycad has a massive trunk.

*Encephalartos concinnus*
The male cones are green.

*Encephalartos ferox*
The red female cones are very showy.

*Encephalartos caffer*
Kaffir Bread. The growth is silver-woolly and crowded.

*E. "Chinhazany".* A variety growing at Fairchild Gardens Florida, with yellow-stalked 1–1.2m (3'–4') fronds, bearing attractive, narrow, blue-green leaflets.
*E. concinnus* (Zimbabwe). Has upright 1–1.5m (3'–5') fronds with a green leafstem and bearing narrow, blue-green leaflets. Cones are yellow-green.
*E. cupidus* (Transvaal). A dwarf grower, on a 15cm (6") trunk which suckers freely into many heads. These bear up to 1m (3') horizontal fronds with narrow toothed blue-green leaflets. Cones are pale green. Sun hardy, light frost-tolerant, good drainage.
*E. cycadifolious* (East Cape). The short trunks freely sucker into many heads. These bear 1m (3') upright yellow-stemmed fronds with deep green, narrow leaflets. Multiple cones are orange-yellow and woolly. Tolerates sun, frost and snow, prefers alkaline soil.
*E. eugene-maraisii* (Transvaal). Trunks to 2.5m (8'), bearing a heavy head of 1–1.5m (3'–5') fronds with close, narrow, blue-green leaflets. Multiple cones have a dark, red-brown appearance. Sun- and frost-tolerant.
*E. ferox* (Mozambique, Natal). The low trunk bears 1–1.5m (3'–5') fronds with holly like, shiny green leaflets. Attractive, bright red multiple cones. Avoid hot sun, winds and frost.
*E. friderici-guilielmi* (East Cape, Transkei). Tall to 4m (12') trunk, topped with stiff 1–1.5m (3'–5') fronds of narrow, dark green leaflets from a woolly crown. Multiple cones are yellow at first, later brown and woolly. Tolerant to sun, frost and snow.

*Encephalartos lanatus*
This specimen is infested with Mealy Bug.

*Encephalartos cycadifolius*
A cycad suited to well-drained alkaline soils and full sun.
(Photo C. Migliaccio)

*Encephalartos hildebrandtii*
Note the reddish female cones, in the centre.

*Encephalartos longifolius*
A lovely and hardy large species.

*Encephalartos inopinus*
An attractively coloured species.

*Encephalartos latifrons*
Spiny Kaffir Bread has vicious leaf spikes.

*Encepharlartos friderici–guilielmi*
This species requires full sun and good drainage.

*E. gratus* (Malawi, Mozambique). Has a heavy 2m x 1m (6′ x 3′) trunk with a thick head of 2m (6′) fronds with deep green leaflets. Cones brownish. Both hardy and sun hardy, but not frost-tolerant.

*E. hildebrandtii* (Tropical East Africa). Large grower, with a trunk to 4m (12′) tall, fronds to 2–3m (6′–10′) and narrow, dark green leaflets. Multiple cones, green then reddish-brown. There is also an *E. Hildebrandtii* var. *dentata* (Tanzania, Uganda).

*E. horridus,* Ferocious Blue Cycad (East Cape). Has short 0.5m (18″–24″) trunks, suckering into a clump, with heads of stiff 1m (3′) fronds, bearing lobed, spiky, silvery, blue-green leaflets. Cone green, overlaid reddish-brown. Hardy, tolerates sun and frost. Needs acid soil.

*E. inopinus,* Spiny Kaffir Bread (Transvaal). Trunk is tall to 3 (10′) and more slender, with 1m (3′) fronds bearing narrow silvery, blue-green leaflets. Multiple, silvery green cones. Hardy, tolerates sun and frost.

*E. lanatus* (Transvaal). Thick 1–1.5m (3′–5′) trunk, with 1m. white woolly, recurved fronds and narrow leaflets; later deep green. Multiple yellow cones, with white wool. Tolerates shade and semi-frost.

*E. latifrons,* Spiny Kaffir Bread (S.E. Cape). Trunk heavy to 3m (10′) with stiff recurved 1–1.5m (3′–5′) fronds, bearing savagely spine-lobed leaves. Multiple, green cones. Semi hardy but no frost.

*E. laurentianus* (Angora, Zaire). The thick 2m (6′) trunk has 1.5–2m (5′–6′) straight fronds of narrow, dark green leaflets, with a few edge spines. Sun hardy, but no frost.

*Encephalartos manikensis*
Another species with attractive long leaves.

*Encephalartos horridus*
Ferocious Blue Cycad. Attractive, but vicious. Note the green cone.

*Encephalartos manikensis*
A large male plant with spent cones.

*Encephalartos munchii*
The leaflets overlap and droop.

*Encephalartos laurentianus*
This cycad has a large, heavy trunk.

*Encephalartos natalensis*
The leaves emerge silver woolly.

*Encephalartos lebomboensis*
Hardy grower, with colourful F cones.

*E. lebomboensis* (Natal, Transvaal, Swaziland, Mozambique). Trunk to 3–4m (10′–12′), with straight fronds to 2m (6′), having dark green, narrow leaflets, with edge spines. Multiple, orange-yellow cones. Sun hardy, but no frost.

*E. lehmannii,* Blue-leafed Cycad (S.E. Cape). The trunk of 1.5–2m (5′–6′) has 1m (3′) recurving, stiff fronds of stiff and sharply pointed, narrow, blue-green leaflets. Cone is green, overlaid with reddish-brown. Tolerates sun or shade and frost. Prefers alkaline soil.

*E. longifolius* (South-East Cape). Sturdy trunk to 4m (12′), with 1–2m (3′–6′) yellow-stalked fronds and narrow, sharply tipped, glossy deep green leaflets. Multiple, greenish-brown cones. Sun hardy, no frost, need acid soil (ph 5.4).

*E. manikensis* (Mozambique, Zimbabwe). Sturdy trunk to 3–4m (10′–12′) and 2m (6′) pale stemmed fronds, with angled, deep green, narrow sharp tipped leaflets. Cones are reddish-brown. Sun hardy, no frost.

*E. munchii* (Mozambique). The trunk is low and sturdy and the fronds 1.5–2m (5′–6′) long and straight on yellowish stems. The silver green leaflets are narrow, with numerous edge spines. Sun hardy, no frost.

*E. natalensis* (Natal). Has a sturdy trunk to 4m (12′) long, and straight 1.5–3m (5′–10′) leaves with narrow deep green leaflets. Multiple cones are golden-yellow. Sun or shade hardy, no frost.

*E. paucidentatus* (Transvaal, Swaziland). On a tall trunk to 6m (20′) with 1.5–2.5m (5′–8′) yellow-

*Encephalartos paudicentatus* (juvenile)
Mature growth resembles that of *E. natalensis.*

*Encephalartos lehmannii*
Blue-leaved cycad, showing a female cone and seeds.

*Encephalartos natalensis*
Grows on a sturdy trunk.

*Encephalartos princeps*
The bluish leaves recurve sharply.

*Encephalartos tangwendte*
This female has two green cones.

stemmed, straight leaves, with narrow, deep green leaflets. Multiple, rusty golden cones. Avoid hot sun, semi-hardy, no frost.
*E. princeps* (South East Cape). Sturdy trunk to 4m (12′) and stiff 1–1.3m (3′–4′) leaves, which recurve at the tips, and narrow, blue-green leaflets angled upwards from the stem. Multiple, olive green cones. Tolerates hot sun and frost.
*E. tangwendte* (origin?) (previously *E.* sp. *"Tangwendte"*, at Fairchild Tropical Gardens). Has a sturdy trunk with 1.5–2m (5′–6′) straight leaves, bearing narrow-toothed, blue-green leaflets. Multiple greenish cones.
*E. tegulaneus* (East Africa). Thick trunk, with straight 1.25–1.75m (4′–6′) leaves, bearing narrow, bright green leaflets with a few edge prickles.
*E. transvenosus*, Mujaji Palm (Transvaal). Claimed as tallest species, with a trunk of 8m (25′) or more. Leaves are straight, 1.5–2.5m (5′–8′) with deep green 2.5–4cm (1″–1½″) wide, tapered, prickled leaves. Multiple, rusty golden cones. Avoid hot sun, cold winds and frost.
*E. trispinosus* (South-East Cape). Trunks 1m (3′) and suckering, with 1–1.25m (3′–4′) recurved leaves, bearing forked and spiked, stiff silvery, blue-green leaflets. Cone is yellow. Tolerates hot sun and frost.
*E. umbeluziensis* (Swaziland, Mozambique). Subterranean trunk, with 1–2m (3′–6′) straight leaves, bearing long narrow, bright green leaflets, few prickles. Multiple green cones. Sun hardy but no frost.
*E. villosus* (E. Cape, Natal, Swaziland, Transvaal). Subterranean clustering trunks, with arching 1.5–3m (5′–10′) leaves, bearing long narrow, dark green leaflets, few prickles. Multiple golden cones. Avoid hot sun and frost.
*E. woodii* (Natal, now extinct in the wild). Largest species, with a heavy trunk to 6m (20′) and a massive head of 2–2.5m (6′–8′) arching leaves, with broad tapering, dark green prickled leaflets. Multiple golden cones. Sun and shade hardy, no frost.
Other recorded species are:
*E. chimanimaniensis* (Mozambique, Zimbabwe)
*E. ghellinckii* (Cape Natal)
*E. heenanii* (Swaziland, Transvaal)
*E. humilis* (Transvaal)
*E. laevifolius* (Swaziland, Transvaal)
*E. ngoyanus* (Swaziland, Zululand, Transvaal and Natal)
*E. marunguensis* (Zaire)
*E. poggei* (Angola, Zaire, Tanzania)
*E. pterogonus* (Mozambique)
*E. schmitzii* (Zaire)
*E. septentrionalis* (Central African Republic, Rwanda) Sudan, Uganda, Zaire)
*E.* sp. A (see Heenan, 1977) (Tanzania)
*E.* sp. B (see Heenan, 1977) (Kenya)
*E.* sp. C (see Heenan, 1977) (Tanzania)
**Culture and habitat** Habitat, and therefore culture, is most varied.
*Encephalartos* species from sand dune areas tend to be smaller (about 1m [3′] high) and have subterranean trunks. Those from more open grassed and wooded areas, on the other hand, may be very sizeable and have heady single or multiple trunks. They store water

in these trunks, so the large-trunked species would be found in drier localities, or those with only seasonal rain. The foliage shows a similar range of shapes and colours, and these features also usually give an indication of hardiness and cultural preferences.

The blue-coloured species will usually stand up to the harsher extremes of heat, sun and drying out. Species with blue-green leaves are slightly less hardy in arid situations than the former; their leaves turning green in conditions where they receive a greater amount of shade and water. Next hardiest are those species with light green leaves, requiring a little more water and perhaps some protection from very hot sun. Finally, the species with softer green leaves require regular watering and protection or filtered shade. These are generally forest understorey plants.

It therefore appears that the blue colouring is associated with hardiness, and if that colouring is maintained by withholding water until it is needed, and by exposing the plant to full sun, the hardiness of the plant will be maximised.

*Encephalartos* species are most showy cycads, and the blue-leaved species make a good contrast with the predominant green of most other plants. The large and showy seed cones are also most intriguing, adding a lot of interest in a sunny garden. The green-leaved species, too, can make a good show in more shaded positions. For instance, *E. ferox*, with its colourful red female cones, stops people in their tracks wherever it is grown. However, do not forget to check what size a species grows to before you

*Encephalartos transvenosus*
Mujaji Palm is the tallest *Encephalartos* species.

*Encephalartos tegulaneus*
This cycad has a sturdy trunk.

*Encephalartos trispinosus*
A hardy cycad with stiff blue leaves.

*Encephalartos umbeluziensis*
A cycad with a solitary subterranean trunk.

plant it. Nor should you forget the spines when selecting the position: always avoid traffic areas. These cycads are showy in cactus gardens, against a fence, or, in the case of the sun-lovers, in a bed in the centre of the lawn.

Those species requiring less intense sun are best grown amongst or under other plants, including palms. All species love a surface mulch of wood shavings or leaf mould, to retain the moisture and keep the roots cool.

**Sex** Dioecious. Needs a male and female for fertile seeds.

**Position indoors** Little is known of their use indoors. They would need to be placed where they could not cause injury, but would still receive plenty of natural light. Do not forget to put them outdoors again to make and harden off their new leaves.

**Propagation** Grow from seeds or offshoots. The fertile seeds (use the water test: fertile ones sink), should be planted, half-covered in potting mix and kept moist. As they show growth, replant them singly. With the offshoots, be careful not to damage their roots, or the parent plant. Dust the cuts with sulphur or a fungicide dust and allow to dry before planting. With stems, also dust and dry off the cuts for a week, then stand or prop upright on sand to let them callus up and make roots. To prevent dehydration, hose the leaves and stem daily. Plant when rooted, but use water sparingly until there is substantial rooting.

*Encephalartos villosus*
This cycad has a large cluster of subterranean trunks.

*Encephalartos woodii*
This massive species is now extinct in its Zululand natural habitat.

# LEPIDOZAMIA (ZAMIACEAE)

Two species recorded, from eastern and north-eastern Australia and growing in rainforests, accompanied by tree ferns.

*L. hopei* (north-eastern Queensland). This species, possibly the tallest of living cycads, appears quite palm-like. It is usually single-trunked, growing 10–20m (30′–60′) high, but branched specimens have been observed. The slender trunk is topped by an attractively arrayed head of smooth, shiny green fronds. The leaflets of this species are broader than those of the other, 17–30mm (¾″–1¼″) in width, 10–32mm (4″–13″) in length, and veined.

*L. peroffskyana,* Scaly Zamia (north-eastern New South Wales and south-eastern Queensland). This cycad has a slender trunk 3–6m (10′–20′) high, topped with attractive arching, shiny green fronds 2–3m (6′–10′) long. The leaflets are 7–14mm (¼″–½″) broad, 10–32cm (4″–13″) long, veined.

**Culture and habitat** *Lepidozamias* grow around the edges of rainforests. The leaf stalks arise from a coarsely, hairy brown base at the top of the slender black trunk. The previous year's fronds hang dead

*Lepidozamia peroffskyana*
Bearing a large female cone.

*Lepidozamia hopei*
This species has broader leaflets than *L. peroffskyana* and its trunk can grow to 3–6m (10′–20′).

*Lepidozamia peroffskyana*
The trunk can grow to 3–6m (10′–20′) high.

*Macrozamia fawcetti*
Inhabits the New South Wales coast and adjacent foothills.

*Macrozamia heteromera*
A small species, ideal for pots, with forked leaflets.

*Macrozamia lucida*
Grows in coastal areas north and south of Brisbane, Queensland.

*Macrozamia moorei*
Large palm-like species from Central Queensland and north-eastern NSW.

after the new set have developed. The male cones, 50–60cm (20″–24″) long, are green and open spirally to release their pollen; a feature which distinguishes these from other cycads. The female cones, which start green and later turn brown, are oval, 60cm (24″) high and 30–40cm (12″–16″) broad, with downturned scales that open to reveal red seeds. These cones are as large as, or larger than those of *Encephalartos*.

10 years ago *Lepidozamias* were not readily available for landscaping in their native Australia and would rarely have been heard of in this trade. Today, however, they are being sold as fast as they can be grown, by chain stores and other large retailers, in little 100mm (4″) pots with only one 100mm (4″) leaf. This will probably be the case with most cycads for many years to come, the world over. *Lepidozamias* prefer a shaded aspect, or one with lightly filtered sun. Full sun may yellow or burn the leaves. They are rainforest understorey plants in nature and these are the type of conditions to copy in landscaping. They would be perfect for ferneries, under tree cover, or in shaded areas between houses and buildings; they are ideal subjects, with palms, for the home rainforest.

**Sex** Dioecious. Needs a male and female for fertile seeds.

**Position indoors** No information is available on their use indoors, but I suggest a position with good natural light. Refer to 'Cycads for indoor decoration' (p. 41).

*Macrozamia communis*
Showing a number of male cones.

*Macrozamia communis*
This specimen displays a trunk.

*Macrozamia communis*
Burrawang, inhabits the New South Wales coast.

*Macrozamia macdonnellii*
Handsome bluish leaves and a trunk sometimes to 3m (10′) high.

**Propagation** Plant seeds to half their depth in potting mix or similar and keep watered well. See also appendix, 'The Cycads of Australia' (p. 13).

## MACROZAMIA (ZAMIACEAE)

17 species recorded, distributed across Australia. Some grow in either flat semiarid country or arid gorges of central Australia. But most grow in eucalypt forests or woodlands in either flat or elevated country, in coastal and inland districts of eastern and south-western Australia.

*M. communis*, Burrawang (New South Wales coast to ranges). The trunk is mostly subterranean, about 60cm (24″) across and to 2m (6′) high. There are up to 50 arching, upright fronds to 2m (6′) long. Plants can carry up to ten male cones or six female cones.

*M. diplomera* (New South Wales inland ranges, in sandy country. This species has a subterranean trunk to 40cm (16″) wide with up to 40 upright fronds on a mature plant with the leaflets forking into two segments at their base and has a resemblance to *M. communis*.

*M. fawcetti* (New South Wales north coast to uplands). This is a smaller species, usually on a subterranean trunk. It resembles *M. pauli-guilielmi*, but broader leaflets and fronds spirally twisted.

*Macrozamia pauli-guilielmi* ssp. *pauli-guilielmi*
Note the spiral twist of the leaf stalk.

*Macrozamia miquelii*
Note the colourful female cone.
(Photo L. Butt)

*Macrozamia macdonnellii*
In natural habitat, Serpentine Gorge, Central Australia.
(Photo D. Stallard)

*Macrozamia miquelii*
A large specimen 2m (6′) high.

*Macrozamia pauli-guilielmi* ssp. *pauli-guilielmi*
A form from the Injune area of Queensland.

*Macrozamia pauli-guilielmi* ssp. *pauli-guilielmi*
Displaying a small trunk.

*Macrozamia riedlei*
*Zamia* Palm. A male specimen in habitat north of Perth, WA.
(Photo D. Stallard).

*Macrozamia secunda*
Small variety with concave leaflets.

*Macrozamia riedlei*
*Zamia* Palm, showing the male cones. (Photo D. Stallard)

*Macrozamia spiralis*
Located in the foothills west of Sydney, NSW.

*M. heteromera* (New South Wales inland and on Dividing Range). This species is found with *M. diplomera,* which it somewhat resembles. It is a small grower, having a 25cm (10″) subterranean trunk with about 10, blue-green fronds. The leaflets frequently divide twice at their base. It generally carries only 1 female cone or up to 4 male cones.

*M. lucida,* Pineapple Zamia (Queensland: Brisbane and areas north and south). This cycad grows to 1.2m (4′) on a subterranean trunk. It has a good display of spreading 1.5–2m (5′–6′) fronds, showing white at the bases of the leaflets. Up to 8 male cones or 2 female cones.

*M. macdonnellii* (Northern Territory: Macdonnell Ranges area). This Central Australian species grows in isolation, surrounded by desert, sometimes with the palm *Livistona mariae.* It is a large grower with a heavy 2–3m (6′–10′) trunk topped with 50 to 100, blue-green fronds. Female cones to 40cm (16″) by 27cm (11″) and nuts 8cm (3″) by 4.5cm (1 ¾″). Sun hardy.

*M. miquelii* (Queensland: lower central coast to uplands). A variable species with a usually subterranean trunk up to 60cm (24″) tall and 50cm (20″) thick, bearing a handsome array of tall, blue-green fronds 2m (6′) long. Female cones are 40cm (16″) long.

*M. moorei* (Parts of central Queensland and northern New South Wales). This is a hugh palm-like cycad

*Macrozamia pauli-guilielmi* ssp. *flexuosa*
This subspecies grows in the uplands of NSW, Hunter Valley.

*Macrozamia pauli-guilielmi* ssp. *plurinervia*
From upland areas of northern NSW into Queensland.

*Macrozamia platyrachis*
Located in a reserve area in Central Queensland. Ideal for pots.

*Macrozamia stenomera*
The leaflets are narrow and repeatedly forked.

*Microcycas calocoma*
This specimen has a trunk 1m (3′) high. A rare cycad.

2–8m (6′–26′) high, with a thick 60–80cm (24″–32″) trunk and a leaf spread of 5m (15′). The blue-green fronds can be up to 1.8m (6′) in length. Each plant carries up to 8 female cones 45–80cm (18″–32″) by 19cm (7½″) or up to 100 male cones, on a large plant. Sun hardy.

*M. pauli-guilielmi* (Eastern Australia) is a species with three subspecies. All are small and compact-growing and excellent as pot plants. These are as follows:

*M. pauli-guilielmi* ssp. *pauli-guilielmi* (South-eastern Queensland). This ssp. bears from 3–10 spirally twisted fronds to about 1m (3′) in length and bearing long narrow, green leaflets, with white bases.

*M. pauli-guilielmi* ssp. *flexuosa,* Kangaroo Nut (New South Wales Central Coast). On this subspecies the fronds may be spirally twisted through 2 complete revolutions. The leaflets are quite narrow.

*M. pauli-guilielmi* ssp. *plurinervia* (northern New South Wales and southern Queensland: ranges). The fronds of this subspecies are stiffer and not as spirally twisted and reddish-brown at their bases.

*M. platyrachis* (Queensland: confined to one inland national park). This is a small species with a subterranean trunk and only 4 to 10 arching fronds. The leaflets are broader and stiffer. The female cones are cylindrical and long.

*M. riedlei,* Zamia Palm (Western Australia: south-west corner). A sturdy species, usually trunkless, but in some areas a very thick trunk of 2m (6′) or more high. A plant may reach 3m (10′) high and 4m (12′) across. The female cones are about 30cm (12″) in length. Sun hardy.

*M. secunda* (Western Slopes of inland New South Wales). A small species, similar to *M. spiralis,* with concave fronds and crowded narrow leaflets.

*M. spiralis* (New South Wales: from Sydney to the Western Slopes). This cycad has a subterranean trunk and about 12 fronds 90cm (3′) long, with pronounced spiral twisting. The female cones are 20cm (8″) long and seeds are orange-red, a deeper colour than in species.

*M. stenomera* (Northern Tablelands of New South Wales). May be mistaken for *M. heteromera,* this species has a subterranean trunk and about 10 yellowish-stalked fronds often spirally twisted. The leaflets are narrow and divide or fork 2–3 times at the base.

**Culture and habitat** With a few exceptions, *Macrozamias* are shade-loving smaller cycads. As with other cycads, the sun-loving species are the giants, having heavier trunks and carrying more and larger fronds. This should be taken into account when choosing a position to plant a species.

*Macrozamias* are not as prickly as many of the cycads, so with the exception of the sun-loving species (see below) they can be merged with other plants into semi-shaded and shaded areas of the garden. They are particularly effective under palms and in the home rainforest, amongst ferns and tree ferns.

The larger sun-loving species are *M. macdonnellii, M. moorei* and *M. riedlei.* These may be used out in sunny areas where a dwarf palm might have otherwise been used, or even in the centre of a group of three smaller growing palms. By the time the palms have reached a reasonable height, the *Macrozamia* will also be noticeable, but it will never outgrow the position as other plants would.

*Macrozamias* are excellent as potted specimens, showing their leaves to full advantage in this way. Use a good potting mix and keep up the water.

**Sex** Dioecious. Needs a male and female for fertile seeds.

**Position indoors** No information known about indoor use. As with cycads, if trying them indoors, select a position of good natural light and do not leave excess water in the saucer. If a plant begins to make new leaves indoors, take it outdoors to a shady veranda or similarly protected area until the new leaves harden up

**Propagation** Grown only from fertile seeds (nuts). These should be half covered in a container of potting mix. When shoots first appear, plant singly to avoid root damage. Keep the plants in a shaded area. Moisture should be maintained. See also appendix, 'The Cycads of Australia' (p. 13).

## MICROCYCAS (ZAMIACEAE)

One species, from Cuba, growing amongst Royal Palms and other vegetation.

*M. calocoma* (Cuba). This very dainty species has a tall stout trunk 3–9m (10′–30′) high, topped with an attractive crown of 1m (3′) bright green, glossy recurving fronds. The leaflets, up to 30 pairs per frond, are closely packed and softer than those of most other cycads.

**Culture and habitat** This attractive cycad was originally reported many years ago, by an early cycad enthusiast on a visit to Cuba, as growing to 9m (30′) tall amongst thousands of Royal Palms. As this particular stand would most likely have been growing in very rich soil, it is doubtful that it survives today. *Microcycas* would be best planted in a semi-shaded position, as very hot sun may burn the attractive foliage. Being tropical, it is probably not suited to cultivation outdoors in colder climates, unless grown in a protected corner, or kept as a potted specimen under protection. It will not tolerate frosts.

*Microcycas calocoma*
The foliage is broad and handsome.

*Stangeria eriopus*
The form from open areas grows bushier.

*Stangeria eriopus*
Landscaped into a rockery garden.

*Zamia acuminata*
A small grower from Central America.

*Microcycas* are rather rare and not seen much away from their native Cuba, even in botanic gardens. However nice specimens were present and growing well in Foster Gardens, Hawaii (tropical) and Fairchild Gardens (subtropical). Both were planted under the light protection of trees.

**Sex** Dioecious. Needs a male and female for fertile seeds.

**Position indoors** No information available on indoor use. Should you wish to try it indoors, select the best natural light and keep moist, emptying any excess water from the saucer. Should any new leaves appear, place the plant outdoors until the new leaves have hardened, in a shady place to avoid sunburn.

**Propagation** Plant fertile seeds (nuts) half covered in potting mix and keep moist. As soon as shoots appear, plant singly to avoid root damage.

## STANGERIA (ZAMIACEAE)

One species with two forms, distributed from East Cape to Natal, South Africa. Grows in coastal grasslands and inland forests.

*S. eriopus*, Hottentot's Head. This cycad has a subterranean stem sometimes branching into as many as 10 heads, each of which may bear a cone. There are about four leaves to a head and the length of these varies between the two forms (one growing in grassland, the other in forest) from 25–30cm (10″–12″) to 1–2m (3′–6′). Each frond carries five to 20 pairs of leaflets. The cones are silvery, becoming brown with age. The male cone is 10–15cm (4″–6″) by 3–5cm (1″–2″) the female 18–20cm (7″–8″) by 8–10cm (3″–4″) and the seeds are red.

**Culture and habitat** *Stangerias* from grasslands tend to be more compact and bushy, while those from shaded forest areas have longer leaf stalks, as can be seen in the pictures. Should you obtain a *Stangeria*, be sure to plant it in conditions appropriate to the particular form you have. For many years this cycad was thought to be a fern, so similar to these plants is it. It was even classified as a fern under the name *Lomaria eriopus* for a long time.

Having non-thorny foliage this cycad does not need to be planted away from traffic areas, so it can be planted in any position that provides its requirements. It requires ample water and both forms do best if they receive some shade. If grown in a pot in a bush house, it tends to retain two or three sets of fronds at any time. In the garden, however, it may tend to become deciduous in more open areas. It will not tolerate frost.

*Zamia floribunda*
Growing at Foster Gardens, Hawaii.

*Zamia fairchildiana*
This small-trunked species has rust-coloured cones.

*Zamia fischeri*
The fronds of this cycad grow to about 45cm (18″).

*Zamia fischeri*
This compact form has shorter fronds and is ideal for pots.

*Zamia furfuracea*
Jamaica Sago Tree is a most attractive plant. Note cones.

*Zamia cremnophylla*
A showy variety growing at Fairchild Gardens, Florida, USA.

**Sex** Dioecious. Requires a male and female for fertile seeds.

**Position indoors** No information available. Refer to 'Cycads for indoor decoration' (p. 41).

**Propagation** Propagation is either from fertile seeds or from offshoots. Sow the seeds into potting mix, half covering them. Water should be maintained. When they begin to shoot, repot singly, being sure not to damage the root. Grow under shade protection, or in a bush house. Offshoots may be removed from an older multiple-trunked plant, but do not disturb the whole plant when taking the offshoot, and be careful with the root section. Pot carefully and grow on in a bush house.

## ZAMIA (ZAMIACEAE)

About 64 species, from the Caribbean Islands and the neighbouring mainland from Florida through Mexico and Central America to northern Brazil. Some grow in sandy coastal scrub and on rocky cliff faces near the sea, while others grow under cover with *Chamaedorea* palms, in tree-covered areas and on rocky hillsides. Sometimes the soils are alkaline.

*Zamia chigua* (syn. *Chigua restrepoi*) (front)
A small, rare cycad from Colombia. The larger species at the rear is *Z. skinneri*.

*Zamia debilis*
This specimen has a female cone.

*Zamia furfuracea* x *loddigesii*
A cross of these two species.

*Zamia furfuracea* (yellow form)
The new foliage is yellow, but turns green with age.
(Photo L. Butt)

*Zamia floridana*
Coontie of Florida, photographed at Lone Pine Key.
(Photo C. Migliaccio).

*Zamia furfuracea*
A larger Jamaican Sago Tree.

*Zamia integrifolia*
Also nice as a pot-plant and bears many orange female cones.

*Z. acuminata* (Nicaragua, Panama). This species has multiple fronds, bearing small, broad tapered mid-green leaflets.

*Z. angustifolia* (Bahamas). Fronds are arching, bearing long, narrow, deep green leaflets. Cones are 12–14cm (5″–5½″) long, reddish-brown.

*Z. chiqua* (syn. *Chiqua restrepoi* a proposed new name) (Colombia). This is a rare species, having a few slender fronds. Leaflets are broad and tapering, with side spines. Cold-sensitive.

*Z. cremnophylla* (Mexico). This zamia has arching, palm-like fronds with shiny bright green leaflets.

*Z. debilis* (Dominican Republic). This species has stalked fronds from the subterranean trunk with long, narrow, dark green leaflets. Cones are dark brown, 12cm (4″–5″) long.

*Z. fairchildiana* (Costa Rica, Panama). A thin trunk bears a few fronds, with notched stalks and broad tapered green leaflets. Cones are brown, 12cm (4″–5″) long.

*Z. fischeri* (Mexico). This species has floppy fronds, bearing small, soft, tapered, serrated green leaflets. Cones are 4–8cm (1½″–3″) long, fawn to brown. There is also a more compact form (see picture).

*Z. floribunda* (at Foster Gardens, Hawaii, but not authenticated from other sources). The fronds of this species have short stalks and glossy bright green tapered leaflets. The validity of this variety is doubtful.

*Z. floridana,* Coontie (southern Florida). Grows on a tuber-like subterranean trunk. The 60cm (24″) fronds with brownish stems are numerous and slender, on this small variety, with bright green narrow concave leaflets.

*Z. furfuracea,* Jamaican Sago Tree (Mexico). This is an eye-catching species to more than 1m (3′) in height, with arching fronds bearing stiffish, almost oval, blue-green leaflets. Cones are rusty brown, 10cm (4″) long. A variant with all yellow juvenile leaves is also pictured.

*Z. furfuracea x integrifolia.* This seed cross has attractive semi-stiff foliage and mid characteristics.

*Z. integrifolia* (Hispaniola) (syn. *Z. pumila* var. *integrifolia and Z. floridana* in some botanical findings). Dense fronds to 75cm (30″) have narrow leaflets. Cones are rust-coloured, 10cm (4″) long.

*Z. latifolia* (Belize, Mexico). The fronds of this species have 30cm (12″) stalks and long slender leaflets. Cones are dark brown, 6–8cm (2½″–3″) long.

*Z. lindenii* (Ecuador, Peru). Fronds are upright, to 1-2m (3′–6′) high, with brown stalks; leaflets are tapered and curved. Female cones are up to 30cm (12″) long.

*Z. loddigesii* (Belize, El Salvador). The fronds have brown stalks and long tapered curved leaflets. The brown cones have a protrusion.

Also *Z. loddigesii* var. *latifolia* (Guatemala, Honduras, Mexico, Venezuela). (Variety status is doubtful, because of much variation in this species).

*Z. manicata* (Brazil). The upright fronds have dark stalks and 20–25mm (¾″–1″) leaflets, mid-green and very shiny.

*Z. muricata* (Colombia, Guatemala). The long fronds have brown stalks; leaflets are deep green, tapered and curved. Cones are brown, 15–20cm (6″–8″) long.

Also *Z. muricata* var. *picta* (Mexico, Venezuela). Not pictured, but similar to *Z. muricata*, only with a leaf variegation.

*Z. portoricensis* (Puerto Rico). Tall fronds with brown stalks bear long narrow blue-green leaflets. The broad dark brown cone is 20cm (8″) long.

*Z. pseudoparasitica* (Costa Rica to Peru). This arched species has attractive feathery fronds which are best displayed in a hanging basket. The bright green, shiny leaflets are 15cm (6″) by 15–20mm (½″–¾″). In nature it is an epiphyte with a 1–3m (3′–10′) trunk and fronds 1–2m (3′–6′) long. Female cones are 40cm (16″) long.

*Z. pumila* (Florida and West Indies) Has two subspecies.

*Z. pumila* ssp. *pumila* (Florida, West Indies). This species has dense fronds with stiff, narrow, deep green leaflets. The rusty cones are 12–15cm (4½′–6″) long. Used in Miami as a hardy seaside border. This species is more usually listed as *Z. pumila.*

*Z. pumila* forma. *umbrosa.* A form of the above, collected by Dr Norstog in Florida, with darker foliage.

*Z. pumila floridana* Perhaps only a Florida form.

*A. pumila* ssp. *pygmaea* (Cuba). This is the smallest of all cycads, having leaf fronds only 6–25cm (2½″–10″) in length. The small leaflets are pale shiny green. Usually listed as *Z. pygmaea.*

*Z. roezlii* (Colombia) This species has upright fronds to 1m (3′) high, with scaly brown stalks and tapered curved leaflets.

*Z. skinneri* (Costa Rica, Panama, Peru). This *Zamia* has a few graceful fronds on arching stalks, with broad tapered, curved leaflets that have a corrugated appearance.

*Z. tuerckheimii* (Belsize, Guatemala). The short, arching heavy fronds have shiny, bright green, broad tapered leaflets.

*Z. wallisii* (Colombia) Fronds are sparse and floppy, bearing few broad, tapering, almost oval leaflets indented with parallel lines.

*Z.* sp. (known as *Z. Hybrid* at nurseries). This seed cross has attractive broad fronds, slightly brown-furry beneath the long leaflets and along the stem. The cones are prominent.

*Z.* sp. (Colombia), at Foster Gardens, Hawaii. This species has many fronds 1–1.5m (3′–5′) tall, with closely packed green leaflets.

*Zamia latifolia*
Attractive in a pot. Note the two dark brown female cones.

*Zamia loddigesii*
A small grower with finely toothed leaflets.

*Zamia manicata*
This Brazilian species has finely toothed leaves.

*Zamia pumila*
Growth is low and spreading. Female cones are rusty brown.

*Zamia pseudoparasitica*
This cycad grows as an epiphyte in nature.

*Zamia pumila*
Used in Miami, Florida, USA, as a hardy border edging.

*Z.* sp. 70.037 (Puerto Rico) at Foster Gardens, Hawaii. The tall 1m (3′) spreading fronds with brown stalks bear a large number of long narrow leaflets. Cones are dark brown, about 15cm (6″) long.

*Z.* sp. 71.848 at Foster Gardens, Hawaii (tropical America). This species has many floppy, long fronds with greenish stalks and long, somewhat limp, bright green leaflets.

*Z.* sp. 82.342 at Foster Gardens, Hawaii. This species, newly planted out, has glossy, bright green leaflets 25mm (10″) wide.

*Z.* sp. (Honduras). This young specimen has not yet developed its mature leaf form. The long, tapering, shiny mid-green leaflets have many spines.

Other *Zamia* species recorded, some confirmed as authentic species, and some awaiting this study are as follows:

*Z. boliviana* (Bolivia, Brazil).
*Z. carpenilla* (origin?)
*Z. commeliniana* (origin?)
*Z. cupatiensis* (Columbia)
*Z. curiaria* (Panama)
*Z. cycadifolia* (*syn. of Z. loddigesii*) (Mexico, El Salvador)
*Z. erosa* (Puerto Rico)
*Z. guggenheimiana* (Cuba)
*Z. inermis* (Mexico)
*Z. jirijirimensis* (Colombia)

*Zamia muricata*
These pictures indicate the trunk, foliage and F cone.

*Zamia lindenii*
Attractive upright growth. The female cone grows to 35cm (14″).

*Zamia portoricensis*
The upright fronds have narrow leaflets and the female cone is dark.

*Zamia pumila* ssp. *pygmaea (juvenile)*
The smallest cycad, known as *Z. pygmaea*

*Zamia tuerckheimii*
The shiny green fronds are attractive.

*Zamia pumila* ssp. *pumila*
A nice clump growing at Fairchild Gardens in Florida, USA.

*Zamia skinneri*
Graceful fronds with corrugated shiny green leaflets on a slim trunk.

*Z. kickxii* (Cuba)
*Z. latifoliolata (Cuba)*
*Z. lawsoniana* (Brazil, Mexico)
*Z. lecointei* (Brazil, Venezuela)
*Z. leiboldii* (Mexico)
*Z. lucayana* (Bahamas)
*Z. madida* (Colombia)
*Z. media* (Hispaniola) (syn. *Z. pomila* ssp *pomila?*)
*Z. merolae* (Mexico)
*Z. montana* (Colombia)
*Z. monticola* (Guatemala, Mexico)
*Z. multifoliolata* (Cuba)
*Z. obidensis* (Brazil)
*Z. obliqua* (Colombia)
*Z. ottonis* (Cuba)
*Z. pallida* (origin?)
*Z. picta* (Mexico)
*Z. poeppigiana* (Colombia, Peru)
*Z. pseudomonticola* (Colombia, Costa Rica, Ecuador, Panama, Peru)
*Z. purpurea* (Mexico)
*Z. silicea* (Cuba)
*Z. spartea* (Guatemala, Mexico), (syn. *Z. loddigesii*)
*Z. sylvatica* (Mexico)
*Z. tenuis* (Honduras)
*Z. ulei* (Brazil, Peru)
*Z. umbrosa* (origin?)
*Z. verschaffeltii* (Mexico)
*Z. wielandii* (Mexico)
*Z.* sp. (Read; Jamaica), (Jamaica)
*Z.* sp. (Read; Cayman Is), (Cayman Is)

*Zamia* sp. (*Z. hybrid*)
This plant is attractive and appears to be a seed cross.

*Zamia* sp. (Colombia) at Foster Gardens, Hawaii.
A tall old plant one of many unidentified *Zamias.*

*Zamia* sp. (Honduras), at Foster Gardens, Hawaii.
A juvenile specimen only available.

*Zamia roezlii*
The growth is upright on a slim trunk.

*Zamia* sp. 82.342, at Foster Gardens, Hawaii.
An attractive new addition recently planted.

*Zamia skinneri*
Growing with the slimmer *Z. chigua* (centre)

*Zamia wallisii*
The light green leaflets have parallel grooves.

**Culture and habitat** *Zamias* are small cycads mostly 1m (3′) or under; only a few species grow taller than this. They generally have subterranean or short trunks; some even grow as epiphytes on trees. They have a variety of foliage types, from stiff and spiky to much softer and lacking spines.

There is little published information on the different species, but the foliage of each species can be used as a guide to its natural habitat and cultural requirements. Those with tough stiff foliage would most likely originate from sunny positions exposed to sea winds, on cliff faces or on sand dunes. They should be hardy in sunny and more open positions, as well as making tough pot-plants.

Those with softer thinner-textured foliage, often growing on small trunks, are likely to originate form more tropical regions, and to require shelter and shade. The more lush their appearance and feel, the more warmth and humidity they are likely to require. In warmer regions, they would prefer a shaded position outdoors; in cooler regions they might be best in a bush house or glasshouse.

When you have decided where to plant your *Zamia*, you may find a little potting mix added to the soil will improve growth, particularly in beds that have not had any soil improvement or added drainage. A surface mulching after planting helps keep the roots cool, as well as retaining moisture during drier periods.

*Zamias* are excellent in pots. Being smaller plants, they often lose their impact when obscured by larger specimens. This is particularly the case when they are bearing their cones, which are produced in the centre of the plant at ground level and may be lost to the eye on those species with a subterranean trunk, which usually have bushier foliage. In pots, the cones are raised closer to eye level and can be seen better.

The cones may be singly or multiply produced, with the female always broader than the male. The cones are small compared to those of other cycads, usually 10–15cm (4″–6″) long and rusty brown in colour.

**Sex** Dioecious. Needs a male and female for fertile seeds.

**Position indoors** There has been limited indoor testing of two species of *Zamia* for indoor use: *Z. floridana* and *Z. furfuracea*. The tests indicated a small amount of leaf loss under poorer natural lighting, but a slight increase in leaf growth under better natural light. The leaves do stretch toward the light, so constant turning or rotation may retain a better overall shape. Allow new leaves to harden out of doors, away from damaging sun.

**Propagation** Propagation is from fertile seeds (nuts) sown half covered in potting mix. Keep moist and pot singly at first signs of shooting, to avoid root damage.

*Zamia* sp. (tropical America), at Foster Gardens, Hawaii.
A large old plant, unidentified.

*Zamia* sp. (Puerto Rico), at Foster Gardens, Hawaii.
Unidentified, but resembling *Z. portoricensis*.

# PALM AND CYCAD LOOK-ALIKES

There are many types of plants that look a bit like palms or cycads and are often mistaken at first glance as a variety. Some of them also have a common name which suggests that they are a palm, e.g., 'Traveller's Palm' or 'Banana Palm'. I have therefore decided to include in this book a small selection of the most easily confused 'look-alikes'. As many readers of garden books may lack the botanical knowledge needed to determine whether a given plant is a palm or cycad. I hope that their inclusion may save such readers frustration.

There are many other plants that could be confused with palms, but space is limited, so this selection will have to suffice.

## CARLUDOVICA

*C. palmata*, Panama Hat Plant (Peru). While this plant closely resembles a palm and is commonly referred to as a palm, it is really a member, not of the palm family *(Arecaceae)*, but of the related family *Cyclanthaceae*. It is a trunkless plant, forming a clump of fan-palm type leaves on 2m (6′) stalks. The red fruits form in a cone-like cluster, somewhat recalling a cycad cone, or a *Monstera* fruit.

*Araucaria heterophylla*
Norfolk Island Pine, or Palm. Ideal in pots and often used as a Christmas tree. If planted in the garden it will grow to 50m (150′) or more.

*Aspidistra elatior*
This specimen is one of the variegated forms. *Aspidistras* date back to grandma's day and were often thought to be palms. For pots or shady garden. Grows to 90cm (3′).

*Black Bamboo*
The stems of bamboo much resemble those of many of the clumping bamboo palms, although the foliage is different. Grows to 2–3m (6′ - 10′) or more.

*Beaucarnea recurvata*
Nolina Palm or Pony Tail. A juvenile pot specimen, much resembles a palm. A larger garden specimen could easily be mistaken for a palm. Grows to 2–3m (6′–10′) or more.

*Musa velutina*
Dwarf Banana (Assam). Many of the banana varieties are ornamental, some with broader leaves than others. Some resemble palms at a glance. Grows to 3m (10′) or more.

*Carludovica* sp.
This resembles many of the palm varieties. It had me tricked when looking through Foster Gardens, Hawaii. I thought it was a *Geonoma* Palm. Grows 1m (3′) or more.

*Blechnum gibbum*
This fern grows on a trunk to 1m (3′) and has a great resemblance to a cycad or a dwarf *Phoenix* palm. It may be grown in a pot or shaded area of the garden.

*Cordyline australis*
Grass Palm or Cabbage Tree. When seen in a large clump or in a pot, this may be mistaken for a palm. Grow in a pot or open garden. Grows to 3m (10′) or more.

*Molineria capitulata* (syn. *Curculigo capitulata*) Palm Grass. This plant resembles juvenile palms such as *Syagrus* and could be mistaken for a clump of germinating palm seeds. Grows to 1–2m (3′–6′).

*Dracaena draco*
Dragon Tree. The aged specimen here has a resemblance to a large clump of palms. The species is very slow-growing and at 5–6m (15′–20′) could be over 50 years old.

*Dracaena marginata*
Madagascan Dragon Tree. This fashionable plant may be seen in pots for indoor decoration as well as outdoors in the warmer gardens and tropics. Aged specimens could reach 5–6m (15′–20′).

*Dracaena reflexa*
(syn. *Pleomele*). A popular indoor pot-plant or warmer garden specimen. It has the dainty appearance of a cluster palm. Grows to about 2m (6′) or more.

*Pandanus cookii*
Pandanus Palm or Walking Tree. This could easily be mistaken for a clump of "Stilt Palms". Grows to 3–5m (10′–15′) or more.

*Pandanus tectorius*
This Pandanus Palm is a common sight along the ocean front. It bears large pineapple-like seed clusters. Grows to 3–5m (10′–15′) or more.

*Ravenala madagascariensis*
Travellers Palm. Not a palm but a member of the banana family. It requires a warm to hot outdoor climate. Grows to 5m (15′) and clumps.

*Cyathea cooperii*
Tree Fern. Large specimens on tall trunks could easily be mistaken for palms or tall cycads. Grows to 5m (15′) unless drawn up by competition of overhead plants.

*Xanthorrhoea glauca*
Grass Tree a common sight in the Australian "bush". Grows on a trunk, with the larger specimens resembling a palm or cycad, on black trunks to 2–3m (6′–10′) high.

*Yucca aloifolia*
This variety 'Marginata' is less likely to be confused with a palm. However, the green variety or *Aloes* may be. *Yuccas* grow to about 2m (6′) or more according to age.

# INDEX

**This index includes both botanical and common names. The number indicated in bold typeface refers to the picture page of each species. Other references are indicated in lighter typeface.**